SAFETY SUCKS!

THE MANIFESTO

Samuel Uriah Goodman

Ian Allison, CSP

A Pale Horse Media Co. Book

www.palehorsemedia.co

Safety Sucks: The Manifesto

Copyright © 2021 by Samuel Uriah Goodman and Ian Allison

All rights reserved. No part of this book may be reproduced or transmitted in any form or by any means without written permission from the authors.

ISBN 9798502118859

Published by Pale Horse Media Co.

Other Books by Pale Horse Media Co.

"Safety Sucks! The BULL $H!* in the Safety Profession They Don't Tell You About." First Edition. S. Goodman 2020

"Obscured: The Pursuit of Radical Self-acceptance." S. Goodman 2020

"Safety Sucks! The BULL $H!* in the Safety Profession They Don't Tell You About." The Expanded Second Edition. S. Goodman 2021

"In His Name." S. Goodman 2021

Dedication

This book is dedicated to Jim Rice. An amazing mentor, a great friend, and an absolutely legendary safety practitioner.

Thank you, Jim.

Introduction

- A word from Sam -

Hello, howdy, and hi! Welcome *to Safety Sucks! The Manifesto*, the second book in the *Safety Sucks!* series. To begin, I would like to thank you for picking up this book. Whether you purchased it for yourself, you were gifted a copy from a friend or coworker, or you "borrowed it" from a colleague's office, I'm glad that it has found its way to you. My hope is that you find value within its pages, that it gives you some ideas on how to make safety suck less, and that it encourages you to go out and *Make the World a Better Place to Work!*

I found myself authoring the original *Safety Sucks! The Bull Shit in the Safety Profession They Don't Tell You About*, after publishing a similar article on

LinkedIn entitled *"the 6 Sucks of the Safety Profession"* in which I reflected back on the struggles I faced over the years as a safety practitioner. The overwhelming number of messages, comments, and emails echoing the same or similar experiences drove me to a couple of epiphanies. First, I realized that I wasn't alone. The "suck" that I have encountered and battled with over the years was not isolated, the "suck" was systemic. I read story after story from those that contacted me, some made me laugh, some made me cry, but they all led me to my second conclusion, something has to change. The stories of frustration, burnout, and depression weighed heavy on my conscious. To be quite frank, they pissed me off! How could this profession, one that seeks to better and enrich the lives of others, a profession that I love, wind up in such a sorry state? Since then, I have dedicated a significant portion of my life to exploring exactly that and fighting like hell to counter the "sucks" of professional safety.

In the original *Safety Sucks! The Bull Shit in the Safety Profession They Don't Tell You About*, we discussed in depth, many of the things that plague our existence as safety practitioners. From our frequent and public beatings for failing to predict and prevent accidents, to the long and often unpaid hours that we are required to work, and everything in between, we laid out an inclusive list of the sucks of the safety profession. In *Safety Sucks! The Manifesto* our intentions are to discuss where these "sucks" stem from and to offer some ideas on how we can make them, at the very least, suck less. We will explore the underlying beliefs that organizations, industries, and safety professionals hold about safety, and how those beliefs impact the role of the practitioner. We will talk about how we can create a more impactful role for the safety practitioner, how the practitioner can be more effective in their day-to-day job, the things that we as safety practitioners should be focusing on, and how we can better our approach to safety overall. More importantly, I want to mention that this book is not meant to be simply taken a prescription, as there are infinite "right ways" to do the same thing. If anything, I hope that you find its content to be thought provoking, that it leads you to go off and explore on your own, and that it provides you with a different lens in which to view the professional practice of occupational safety and health. Ultimately, my hope is that you find it valuable, useful, practical, and intriguing.

I want to take a moment to acknowledge my coauthor, Ian Allison. Without his help this book would not have been possible. There are certain people that you meet in the course of your life that are just on another level completely, Ian is one of those people. His level of care, know-how, knowledge, and "give a damn," is inspiring and he serves as an example to all those that he encounters. Ian is more than a colleague, he is a great friend, a friend that I am thankful for on a daily basis. When working together, we just seem to have this way of extracting thoughts and ideas from one another. Through our dialogue, debates, and friendly arguments, *Safety Sucks! The Manifesto* was born.

I hope that you walk away from this experience with some ideas of things you want to change, some thoughts on how you're going to change them, and on a mission to make safety suck less. I hope that it inspires you to seek and destroy the "suck" that has long plagued our profession. The professional practice of safety, even in its current sad state, is far too valuable to set out on the curb. We have a responsibility to leave this profession better than how we found it, we must fix what is broken. Together, and only together, can we accomplish that goal. Safety sucks, but it shouldn't!

We really do have one of the most bad ass jobs on the planet. It's about time we stop accepting things as they have always been, and purposefully excise the "suck" from our profession. As much as I've enjoyed the experience of writing these two books, I long for a better time. I can't wait to be able to write *Safety Sucked!* rather than *Safety Sucks!*

Sam Goodman – The HOP Nerd

- *A word from Ian* -

Yá'át'ééh! I'm Ian Allison and thrilled to be a part of this project. My greeting was in Navajo, I'm born and raised in Arizona on the Navajo reservation, aka "The Rez." Specifically, Tuba City, Arizona. As a Rez kid I never envisioned myself being an author because that just never seemed like a possibility. This is my first time writing anything that is published. I'm so excited to bring this project to you all because it is something I'm passionate about, making the world suck less than it did yesterday. First, a little bit about myself and how I fell into this crazy occupation we know as Safety.

As I mentioned, I'm Navajo and from the Rez. What you don't know is I'm a nerd, I've always been a good student. Student in the classroom and in life. I have two constants I try to keep in my life. Never stop learning and always be willing to say, "I don't know." These have afforded me great opportunities in my career and life thus far. I can confidently say the only period in my life that went as I intended was where I attended college.

I've been fortunate in my life to have gone to school twice after college. First after high school I attended Dartmouth College and was an Environmental Studies and Native American Studies double major. Following my undergrad, I began my career in safety and along the way I realized that business school would be good for my career. So, I graduated from Arizona State University with my Masters in Business Administration (MBA). As a result, I brought as much business knowledge and practical application as I could to this book. Frankly, that is something that most safety professionals don't have enough of, business acumen. I wanted to bring that knowledge and context to our ideas because that's what we miss so much in our profession...how to make our solutions practical and palatable for leadership. Basically, this meant that I would ground Sam's wild ideas with business fundamentals, and he would encourage me to be more daring in my thinking while pushing the envelope. We were a great team throughout this project, and I look forward to working with him in the future. So, let's round this out with how I got into safety and where I am today.

I began my Safety career right after I got my bachelor's degree. I was jobless in 2010 and stumbled into an entry level operations and maintenance program at a power plant near my hometown. After initial training, I briefly swung a sledgehammer for a living as a railroad track laborer and then jumped ship to be closer to what I sought after, an Environmental job at the plant. My undergrad degree was in the Environmental field and it made sense that Safety was a bridge for that to happen. However, I never made my way over to that field, I stayed in Safety. For the next ten years I stayed in Safety, spending most of my time working at power plants, coal and gas plants specifically. I recently moved into Supply Chain management. Frankly, just to see if I wanted to do something else other than Safety. I never sought out to do it, but I just stayed in, got my CSP eventually and never looked back. Now that I've worked outside of safety for a few years I realize that you can influence safety from everywhere. It isn't a hat that you take off or a title or position. While my title no longer has Safety in it, I live this and I'm going to find my way back soon. As much as I enjoy branching out, Safety is where my passion lies. Which is why when approached with partnering with Sam on this book, I couldn't pass it up.

I've known Sam for six years. We met when he replaced me in one of my positions. I had never met him before but during the turnover in the position, I realized he was different. Not the typical safety professional that I was accustomed to dealing with but also, he was YOUNG! Finally, someone who wasn't old and crusty, no offense to those who are old and crusty. I was in my late 20's and Sam was the same. I had never met someone in safety, up to that point, that was even a decade away from me. So, it was refreshing to meet another younger safety pro, but also to meet someone who was an anarchist. He had not yet written *Safety Sucks!* but I could see he had a lot of ideas and opinions about the current state of safety. So, I glommed onto him and we formed a great friendship as we worked closely together at work and outside of it.

Sam came to me with this crazy idea of writing a book, upon us spending so much time just riffing on different safety topics. It began as I was his first in person guest on *The Hop Nerd* podcast and we found ourselves continuing to chat for hours after each episode. So, we decided to just make it official and create something together. As Sam mentioned above, we have a way of challenging each other's ideas and cultivating ideas. Some days in our brainstorming sessions I swear we have the cure to the world's problems, so we decided...let's write it down! What started as just kicking around ideas, Sam told me he had written some stuff down already. When he showed me what he had, he offered for me to go in with him. I couldn't turn it down because he had so much passion around it and I'm so glad I did. Sam is a great friend, motivator, and just overall a great human being. He is one of the most impressive people I've ever met in this world. I was honored to work with him on this. I hope you all enjoy the book; we had a blast writing it.

Ian Allison, CSP

Safety Sucks!

The professional practice of occupational safety and health has within it the ability to help and the capacity to harm. The safety practitioner, the one that wields this doubled-edged sword, holds within their hands a disease and a cure. Unfortunately, the disease seems systemic and the cure far, far away. Our dated and flawed approaches to worker safety have left much to be desired, both within our own ranks and the companies that we serve. While many of these methods have yielded results and some could even be considered foundational to occupational safety, have they reached their peak? Yes, we have achieved the point of saturation, we have squeezed all that we can squeeze from doing traditional safety harder. Unfortunately, we have also discovered the point in the *dose-response relationship* in which traditional approaches to worker safety have become deleterious. Yet we continue to double down on doing the same things harder, all the while our results have diminished, and our organizational cultures degrade. While this book is not meant to argue the effectiveness of traditional safety v. *Safety Differently*, many of the "sucks" within our profession derive from frustrations with evermore useless traditional approaches. Additionally, the way that safety practitioners are frequently viewed and treated by companies, frontline employees, industries, and even other safety professionals, is rooted in the same or similar traditional beliefs around safety.

As safety practitioners, we are often asked or demanded to rid our organizations of the evils of risk and to ensure that bad things never happen. We are tasked with predicting and manipulating the unknown in an everchanging and complex world. We are tasked with creating perfect order from imperfect chaos, to turn all that we touch from complex mayhem into perfect linear orderliness. Safety practitioners are doomed to utterly fail, right from the start. Where does this come from? Why do our organizations believe this to be true? Why do we as safety professionals fall for this "myth of the safety professional?" Because it supports our underlying and deeply rooted beliefs about safety at work in general. There are a few key noteworthy "sacred cows" of traditional safety that are the likely culprits:

> *All incidents are preventable*
>
> *Closely examining and preventing small events allows us to predict and prevent big events on the horizon*
>
> *If people just followed the rules, nothing bad would happen*

When shit inevitability finds a fan, rather than honestly reflecting on the effectiveness of these deeply rooted beliefs, we double down on doing them harder. We write more rules, we preach to the frontline a sermon about caring more, we measure and incentivize more, we beat the involved employees for not following the rules hard enough, we hold safety professionals accountable for failing to predict and prevent, and we try to blame and shame our way to safety success. As practitioners, companies frequently demand that we believe in, preach about, and promote these sacred cows that many of us believe to be untrue. With this conflict, is it any wonder that we struggle to connect with our frontline employees and that we fail to impact those above us? Is it any surprise that we can't seem to make any meaningful positive change in our workplaces? Is it really that shocking that safety practitioners flee this profession due to mistreatment, abuse, and burnout? It's not that surprising at all, in fact, it's become an expectation for safety practitioners and their employers alike. We continue to cling to these strange beliefs that we get better by doing the same things harder, if we finally rid our companies of bumps and scrapes then we will stop killing people, the more the safety practitioner suffers the less others will, and if we finally fix people, all will be well.

We find ourselves in a position in which we love what our job stands for, but we hate what it has become. We have a deeply rooted desire to help people, to make workplaces better, and to put an end to life altering work-related events, but the collectively held beliefs about workplace safety, those held by organizations, industries, and sometimes our own profession, prevents us from doing just that. Often, our input is only valued if it aligns with the most sacred of traditional safety beliefs. We are quickly shunned and labeled as "not caring" if we speak out openly against those sacred safety cows. We are told to care, but to only care in the agreed upon and prescribed ways. We are forced to live and breathe the cult of safety, it is demanded that we preach from a bible in which we do not believe, and that we do it all with a smile, while ruffling no feathers or challenging any of what the organization or our profession has deemed to be good and just.

Humanity has a long history of doing wildly dumb things for extremely long periods of time. As humans, we once believed that smoking had health benefits, not only in its typical method of consumption, but in the application of tobacco smoke enemas that were prescribed regularly by doctors in the 18th century. It was once a commonly held belief that mercury and its "magical healing properties," was a cure all for sexually transmitted diseases and could potentially grant a person immortality. We once collectively agreed as humans that disease was spread through foul smells, and that *germ theory* was a load of bull shit! Benzene was once a common ingredient in cologne because of its "pleasant aroma," we used to believe that if people traveled over 30 mph that they would suffocate to death, that lead and asbestos should be put in practically anything and everything, and on, and on. While it's easy to look back and chuckle at these now comical old practices and beliefs, humanity is great at taking turns into the darker side of who we are. Often, large groups of generally intelligent people, find themselves doing vile and evil things due to the opinions held by the majority. We once believed that it was perfectly acceptable to enslave entire races of people, that performing mass-scale murder against those that we do not like or agree with was good and just, and those are just two examples on a long and ever-growing vile list of human atrocities.

Now, what if we applied the same logic that we apply to safety to these situations? To begin, our conversation would probably never take place. It would immediately be shut down with something like this, "How dare you question the healing properties of smoking tobacco! You must not care as much about people's health as I do!" Now, assuming that the discussion continued, we would more than likely hear something like this, "But that's the way we have always done it! We have been doing it this way for a hundred years!" When those that rely on the "healing properties" of tobacco ultimately succumb to their pre-existing illness or the tobacco itself, we'd then double down on its use. "We must have just not blown the smoke hard enough up their asses!" or "We just need more awareness around the benefits of tobacco smoke enemas," we'd exclaim. As ridiculous as it sounds, replace the good old tobacco smoke enema with safety. Not so funny now, huh? If we approached medicinal tobacco smoke and mercury in the same fashion that we currently approach safety, they would still be widely in use and medical practitioners would be touting their superiority for using such tried and true methods. Tobacco smoke enema anyone?

So, what's the point? We are just as great, if not better, at doing wildly dumb safety things for extremely long periods of time. Upon challenge, we defend these catastrophic beliefs by axing off the heads of those that speak out against them, all while sitting upon our moral high horse virtue signaling. We then persist in our beliefs that if people only cared more, tried more, or did traditional safety harder, then all would finally be well. The results of clinging to our old and tired ways have been horrible for both our profession, and those that are supposed to benefit from our existence. But hey, at least we all got together and agreed that this was the correct way, even if it is dumb. Ultimately, the point is this, simply because the majority holds an opinion, it doesn't make it true. Progress and betterment hide within dissent and minority opinion. If we really want to create positive change within our profession, if we want to make our profession more effective and beneficial to those that we serve, and if we want to make safety suck less, we must openly admit that, just because we've done things this way for a really long time, doesn't make them right.

Change is incredibly hard, even more so for a profession that is well-known for being set in its stuffy and starchy old ways. But we move beyond bad ideas by introducing better ideas. The dated and disproven ideas that yield negative results for us as professionals and cause harm to those that they were intended to help, they will continue to persist until we offer something better. Bad ideas die in open debate; conversations about better ways to do things prompt further discussion and debate. As tiring as it can be, if our desire is to improve the professional practice of safety, and to ultimately better the lives of those that we seek to help, it will be a long and hard battle against those deeply entrenched sacred safety cows and beliefs.

Our profession and the diverse, talented, and knowledgeable people that make up its ranks, deserve much better. The people that our professions serve, do as well. The normal, everyday things that we have grown to accept as safety practitioners and as companies that employ safety practitioners, the post-accident beatings, the expectation that practitioners be safety miracle workers, the belief that safety professionals work best as safety enforcers, the idea that magical safety people can and must predict and prevent every bad thing from happening, that they should be gurus and safety priests, and all of the other horrid things that stem from our current underlying beliefs about safety and the professional safety practitioner, should have gone the way of tobacco smoke enemas and medicinal arsenic decades ago. But yet, as with traditional approaches to worker safety, they seem to only be growing in use and popularity. But if we're lucky, pretty soon, we'll evolve to trepanning people's skulls to remove evil safety spirits from their heads.

Why? Why do we persist in these beliefs even though we know that they are harmful? Simply put, it's two basic, but enormously powerful motivators:

> *On the surface, it all appears morally sound*
>
> *It's super easy*

If we say something like, "no one should get hurt at work," that's a pretty morally sound statement, you'll get no argument from me. Now, when most of us hear a statement like that, we typically think something like this, "Yes, absolutely! No one should suffer a life altering event or die while simply trying to make a living." But that's usually not what this seemingly positive statement means. It is often taken to the extreme meaning that says, "We shouldn't even have bumps, bruises, or scratches! We must achieve absolute zero or we suck at safety!" Why? All roads lead back to the underlying beliefs about what safety is, and how we define that. If we genuinely believe that, a.) *All incidents are preventable* and, b.) *we can prevent catastrophic events by managing and preventing lower-level events*, then that all makes great sense. Unfortunately, life is not that easy. Although both of these key points make up the bedrock of most traditional safety management programs, they are near fairytale like illusions that only lead us astray and farther away from a focus on what actually matters. But, if taken at face value and not examined below the surface, it is easy to see how we lean into them as morally superior.

As people and as companies, we're in love with easy. Blaming your local safety practitioner for failing to predict and prevent an event is much easier than investing time, money, and other resources into learning about and fixing things that are buried deep within our organizations. Blaming and shaming that pesky employee, that unfortunate soul that found themselves at the end of a long line of problems, is much easier than deep reflection on the context that surrounded whatever bad thing happened to them. Telling people to care more, demanding greater situational awareness, blame, punishment, sticks, and carrots, all easy, easy, easy! Demanding that safety practitioners be held accountable when all of that "easy button" safety garbage fails, is just as easy. But, as John F. Kennedy so famously said in 1962 as he rallied support for America's mission to the moon, *"We choose to go to the moon. We choose to go to the moon in this decade and do the other things, not because they are easy, but because they are hard…"* *JFK's* powerful statement acknowledges that growth is born out of vision and struggle, that we often must choose a path that others do not if we seek innovation, and that easy is rarely, if ever the way to achieve greatness.

While I could go on, and on about the sucks that reside within the professional practice of safety, I won't. That "suck" has already been captured in the original *Safety Sucks: The Bull Shit in the Safety Profession They Don't Tell You About*. I felt it more important to discuss where much of this "suck" comes from, rather than use this as a recap of all that ails us as safety practitioners. I do not have to describe to you all that is less than desirable in your job, we live within this profession together. Let's level, safety does suck, but it doesn't have to. We work within one of the coolest and interesting fields that exists; safety shouldn't suck! But yet, due to all that we have touched on so far, safety does indeed suck! With that being said, we must acknowledge that we choose to make safety suck less, not because it is easy, but because it is hard… We have a long road of hard work, innovation, and struggle ahead of us. We have many minds to change, and we will continue to battle with traditional mindsets from which those beliefs grow for many moons to come. But every ounce of struggle and every inch of gained ground, it is worth the fight. The professional practice of safety, even in its current sad state, is far too valuable to set out on the curb. We must fight for what our profession stands for, for enriching and bettering the lives of all those that we touch. We must create a safety profession that affords us the opportunity to *Make the World a Better Place to Work*. Anything other than that, is not worth our time.

We Need to Talk...

We need to talk about why it's so hard to talk about safety. As organizations we love to talk about how to direct safety at people. This comes in the form of programs, recurring meetings, rules, regulation enforcement, and company demands. I bring these items up because when it comes to modifying one of those aforementioned items, we LOVE to talk about safety. Talking about safety in those arenas makes us feel like we are making an impact. A discussion about programs, rules, or actions is good because we'll leave the room with something to, "Fight the good fight." Also, it makes us feel like we are pushing the rock up the hill. We are struggling; therefore, we are growing. Ultimately this is exemplified in the idea that we need to pursue something tangible so in the end we can look back and say, "Look what we did!'"

Now, this critique is not meant to stifle the progress of an organization. More importantly, not to belittle any contributions to a maturing organization. I'm not specifically referring to organizations who recognize that safety is not something you do to employees. It is not a direction or action that is to be applied to an existing culture, rather it is the foundation that propels your cultural change. I am looking at organizations that are improving their company culture. Actively working on their continuous improvement and listening to their people for solutions. Those companies, we don't need to talk. For everyone else, we need to talk because this is about to get real.

For companies that fall into this other category, we mean an organization that views safety as a bolt-on of existing culture. Something that can be improved overnight, or worse something that can be controlled by care and awareness, "If they just were more careful...this wouldn't have happened." Yeah, that's the kind of organization I'm talking about. Why dedicate an entire chapter to just talking? Well because in the year 2021, there are just topics that are too difficult to talk about, for example race, gender, politics, COVID-19, etc. Depending on the tone, perspectives, intent, curiosity level, and audience, a constructive conversation can't be held around these topics. Safety unfortunately has been one of those topics, but not in the same way as the previous topics mentioned.

Why care about safety? Really, ask yourself that question, quietly to yourself outloud say it, "Why do we care about safety? What is the purpose of safety?" Did that question make you feel uncomfortable to ask? Well, it reveals a bias and preconceived notion around questioning safety, it potentially challenges the care that you demonstrate for people. Not in a way that measures your level of care against others, but in a way that challenges your own capacity to care about employees. At the root of our difficult conversations we have about safety programs, reactions to incidents, safety policing, discipline, etc., it comes from a place of care but we also need to recognize that we need to be able to challenge everything. Safety programs seemingly are etched in stone, progress is impossible because under the subpar program is an unwillingness to let go of something that needs scrutiny. These subpar programs are usually hidden behind the veil of "It's the law," regulatory overcompliance, fear of audit, or worse because, "The person who made the rule will get mad if they hear us talking about changing it."

In addition to getting people mad about safety, we refuse to have hard conversations about safety because we don't know what we want. In any organization you'll go in and see signs and messaging that point to zero injuries. *"Return home the same way you came, with ten fingers and ten toes."* Pushing that message is frustrating because it pushes a culture of zero, but also promotes a culture of suppressing the truth. This same organization will have the audacity to be surprised when a fatality occurs because, *"There was no way we saw this coming!"* Frankly, most organizations come across fatalities this way, why? Well, you know the saying, A rising tide raises all ships? Same goes for a low tide, it lowers all ships. We unfortunately are willing to sit in the same dock with other ships because it feels safe. If everyone lowers and bottoms out on the sea bottom, it's okay because everyone else is failing at the same rate as us too. I understand I went a little too deep with that analogy but you get the idea. Benchmarking in safety is bad, yeah, I said it.

Where does the idea of benchmarking come from? Let's start there, well I can tell you that it didn't start in safety. Take a step back and think, why benchmark? Well, in any industry you are competing against your peers for market share and advantage. Any actions that you take that can increase your revenue, lower your costs, or give you an advantage to gain more market share, is a threat to your peers. So you benchmark, to ensure that you can see where you are in relation to your peers. From here you can match up your key performance financial measures like, return on equity, operating cash flow, net profit margin, etc. You know where all of these can be found, for a publicly traded company? On their balance sheet, income statement, and statement of cash flows. These provide you all the information you need, at a high level, to see how you match up with your peers. However, you can take it a step further and measure ways that money is created like productivity, employee size, compensation, fringe benefits, facility costs, equipment costs, etc. This is where it gets messy because we also include safety in this "Step further."

    Did you know that all publicly traded companies need to provide all of their financial data to the Securities and Exchange Commission? It's the law, required so that investors can decide if they want to put their money into this company or not. A 10-k is an exhaustive packet produced annually that ranges from 100-300 pages, typically. In this book of information is data about financial performance, industry headwinds that the company is keeping an eye on, current challenges, future challenges, and all the financial data you could want. You know what is RARELY ever mentioned in this 10-k? Safety. Prior to 2020 safety was only mentioned to reference collaboration with a regulatory agency or as an expense on the balance sheet. However, due to COVID-19, 10-k's for a number of companies had more mentions about employee safety because of the increased costs and challenges it took to maintain operations. So, why bring this up? Well, when we are looking at information that we boast about to investors, safety isn't on there. However, when we talk about safety internally it's paramount, in some companies. Why bring this up? Well because when it comes to financial performance and benchmarking, comparing stock performance against our peers we need to measure ourselves. Looking at the balance sheet, income statement, and cash flows of a company can help you assess it, but really what you want to do is benchmark on things like compensation, training, fringe benefits, worker productivity, employee turnover, etc. Somewhere along the way we started including safety in this process.

You remember years ago when safety professionals all tried to sound like they were new age thinkers and said that lagging indicators only tell you where you've been and not where you are going? We needed to look forward. *You can't drive your car by looking at your rearview mirror, you need to see what's ahead of you!* Frankly, safety just piggy backed on what the industry was doing with their comparative analysis. Companies realized that strictly looking at financial performance at one moment just gave them a snapshot in time, they needed numbers that told them what might be coming and how they can plan accordingly, leading indicators. You really don't think that safety came up with leading indicators, do you? Good catches, close calls, actionable close calls, etc. No, these are all ideologies that businesses used to gain a competitive advantage and lower their risk, so they applied it to their existing metrics in hopes that it would improve their performance. So, what happens when you attempt to measure safety like it's going to directly impact your bottom line? You get mediocrity masked as innovation. This is the difficult conversation that needs to be had with organizations, stop measuring safety like it's something that can be measured and compared against your peers.

The best companies are good at two of the three things: Price, Quality, and Service. They are exceptional at the remaining of the three. Walmart, okay on quality and service but excellent on price. Amazon, average on quality and price but exceptional on service. Tesla, service and price, in relation to the electric car market, are okay, but quality is exceptional. How does this tie into safety? Great question, it doesn't. Safety is everywhere and nowhere at the same time. The culture that you have dictates the safety of your company. Do you believe that good ideas can come from anywhere and anybody? Chances are you have a healthy relationship around safety? Does your company have a rule for everything and all the rules must be followed? You probably have blinders on about what your safety performance truly is, but I wouldn't doubt that disillusion doesn't exist in other areas of your company as well.

Safety conversations get muddied by discussion about care and awareness. "*If only they cared more...If only they were more aware of their surroundings...if only they said something, this might not have happened.*" Frankly that is ridiculous. Why? It proves that we have no idea what to do with safety! We treat safety like it is a key performance indicator (KPI) that will impact the bottom line and take investors to the promise land. Simultaneously we treat safety like it's a sacred cow that we can't alter because "*if it ain't broke don't fix it.*" While at the same time we push our leaders and safety professionals to be innovative and creative with how we "Do safety," and see what our peers in the industry are doing. You know what we end up with? An entire industry that has no idea what they want out of safety. Let's start with companies who think they are innovative.

Outside of safety what does innovation look like? Well, it starts with KPIs, setting ambitious goals and using metrics to track your progress to ensure you are on your way to hitting your goal. You know what we do in safety to set an ambitious goal? Let's set the number to zero (0)...again. We are chasing a ceiling. But you know what zero really means for a company, this is where benchmarking comes back in. "I want to have zero injuries that's the goal, I really want to send everyone home the same way they came in. BUT if you are going to get hurt, just as long as I do better than my competitors and peer companies, I'm okay with it."

You can't have negative injuries but we chase zero injuries and are satisfied when we don't hit zero because as long as that number is lower than someone else, we're okay. We have this *"we don't have to be the fastest, we just have to outrun the slowest person"* mentality. This mentality wouldn't be a problem if most companies didn't have this mindset about safety. Truly when you strip away the slogans and "Brother's keeper" signs, they really are saying it's okay if we don't hit zero because we just need to not hurt people at the same rate as our peers. As long as we aren't last or near last, we're good. We can sleep good at night knowing that we are hurting our employees at a rate that is better than everyone else, so it's okay...but we'll keep pushing zero. What is wrong with us!?

You know what chasing zero gets us? What it truly gets us is a false sense of security. Having a goal of zero masks any learning, you don't get a true representation of where you are because by virtue of having a ceiling as a goal, you will hit that goal or get pretty darn close to it. Why? Well because you have employees that work for you and care about keeping their jobs or care about keeping the people who sign their paychecks satisfied. Let's frame it this way, if your goal every year was to increase profit by 5% over the previous year, it was the number one goal of the company. The number one mission, no matter what. Guess what, it would happen every year. Even if it wasn't entirely accurate. If that was your only goal for success, why wouldn't you find all the possible ways to manipulate numbers, reframe data, or justify not meeting that goal. Honestly, some fraud and fudging of numbers would eventually creep in because we have to meet the goal, and we'll do whatever is necessary to keep our jobs. The same thing occurs in safety, but we don't acknowledge it as fraud or cheating or stealing, but we should.

An incident, regardless of what it is, is company property. The company owns that busted knuckle, rolled ankle, shattered hip, and fatality. Not own in the sense that they own the employees that work for them, no, they own what happens to their employees because they're doing a job for the company. I'm not saying that reporting of incidents is going to cure everything, I'm not saying it will even help. But I can tell you that withholding information is a crime when it comes to finance, law, regulatory compliance, etc. So why isn't it for safety? Well because we don't know what to do about safety.

I really don't believe that all incidents have to be reported, but I do believe that incidents belong to the company, in the same sense that values belong to the company, and we need to be keeping account for what's happening. Where we go wrong is thinking that we can measure success out of those incidents. We can glean some insight into the future about what will happen and when. Unfortunately, if it was that easy, it would have been done already.

What we need to start doing is asking ourselves, what are we trying to measure? Really what are we measuring with safety, not just focusing on zero. But truly what are we trying achieve with all of the tracking we are doing. We want to not kill someone right? So, let's throw away all the metrics and measurements that don't help us not kill people. Observations program? Get rid of it. Good catches and close calls? Get rid of them. OSHA recordable injuries, track them because you're required by law but don't publicize every injury like it's a tornado warning for the community. The reason being, outside of OSHA recordable injuries, the other measurements are there to "Track and trend." Just putting those words on paper make me want to gag. This is where we get into leading indicators and lagging indicators, back to us not knowing what to do with safety. We treat safety like a financial metric that we can anticipate headwinds and brace ourselves for changing market conditions when we have leading indicators. Leading indicators can give us insight into how resilient we may be, how spread thin we may be, also where we are tracking to be in the future. These are great for financial, bottom line numbers, but not for safety. Once we stop treating safety like a bottom-line metric, we can begin to have productive conversations about how to make a difference. So, let's say your company decides to get their head out of their ass, what does a positive conversation look like? An answer can be found in how we choose to be innovative and how we should be thinking about how to be innovative.

What does innovative look like? Well, I can tell you that it is best executed when it comes from top-down management that is executed effectively at every chain of the organization. It also looks like organic change driven from the frontline that flows up to management and garners support and translates to other parts of the business. What this means is it looks like a group of employees flustered about goals being handed down by upper management and finding a way to achieve the goals, despite how insurmountable they may seem. It also looks like employees coming up with an idea that is presented to management and eventually supported because we want to achieve a goal here.
Innovation in safety unfortunately looks like setting more goals to bring us to zero injuries and/or adding more programs to our existing library of safety programs. Innovation should be met with vision; we are not innovative in safety. Our vision is zero, so we innovate different ways to get to zero and a company that claims to be innovative while chasing zero injuries is an oxymoron. Those two things do not exist in the same reality, why? Well let's look at how employees receive innovative news in a zero culture. "Team, we're

going to rollout a new metric. Actionable close calls. This metric will help us ensure that when we have close calls, we aren't just reporting them and forgetting about them. We want to make sure we are taking necessary action to reduce likelihood of that close call becoming something serious." No employee listening to this rollout is thinking, "Wow they really care about us, I'm so excited to see how this is going to improve things." No! What they are thinking is how this is going to make their job suck more than it did yesterday. However, we get sold on bullshit metrics, initiatives, programs, off the shelf safety management, etc., to the point that we pass off those ideas as gospel. Ultimately, they fall flat because they were solutions looking for a problem. Does that sound like innovative thinking? No but that's what passes for innovative thinking in safety. Do you want to know how you're onto something? The employees will tell you.

Actual innovation in safety should feel like a breath of fresh air for the employees. Let's say a plant manager comes to an all hands meeting and levels with everyone, "Look, I know we've been pushing zero injuries for a long time at this site. I have firmly believed for a long time that was the right path for us to take, it's not. Why? Because if we hit zero or not, I have no idea where my next fatality is going to come from. That terrifies me, it keeps me up at night. The work that you all do is dangerous, it can alter your life or kill you all together. Of course, I'm not telling you something you don't already know. So as of today, no more tracking of leading indicators. All metrics will be tracked and discussed only among myself and the safety professional. If we feel something needs to be brought up, we will do just that. However, no more publicizing where we stand in relation to last year or compared to our best year we've ever had. In fact, we are stripping away all safety programs and procedures. I'm working with the safety professional to maintain a minimum level of compliance, why are we doing this? Don't we always say compliance is the minimum we strive for? Well up to this point, we've prioritized compliance so much that I'm worried I don't allow you to be creative in your positions and get work done. This is to establish a collaborative environment but also to tell you to take control of your work again. You are the expert, not us. We want to get out of the way and have you tell us how the job needs to be done. We're here to let you know when you're out of bounds, but you're playing in the game not us. So go do it." So, the loud thud heard after the plant manager is done speaking is from the jaws of everyone in the room, collectively dropping. Telling your employees that you were wrong about how you managed safety? Wow. Telling your employees that they are the experts at their jobs and we're going to take a step back to let you do that? Wow. Finally, the proof that this is innovative, it's a vision of trying to not kill people at work by empowering the employees to take the lead on combatting this ever-present threat. But this is blasphemy at any organization because it takes control away from management. There's potential for regulatory non-compliance, and worse what's going to happen!?

The employees embracing and welcoming the change is something we all wish would happen. Frankly even if you addressed your employees at an all hands meeting and said everything verbatim, no one would believe you. Employees would need to see demonstrated commitment to their actions before they would even begin to adopt what was rolled out. Innovation in the workforce looks like "oh gosh how the heck are we doing to get all of this work done!?" But that thought is also coupled with the idea that the vision is attainable and worth pursuing. The vision most places have of lowering injuries to zero only ensures you're going to get to zero eventually. Either for a short duration or through fraud, non-reporting of injuries. So, if zero is such a bad idea why does everyone track it? How can an entire industry be off track? Great questions, let's talk about why zero doesn't help you when you're trying to stop killing people.

Ultimately, we track zero because it feels like a good goal and because our peers do the same. Bottom line, when we compare ourselves against our peers and have that dictate how we manage our employees, we are committing to killing employees at the same rate as our peers. Yeah, we might all be hitting around the anticipated number of injuries every year and track the same with serious injuries, but the ones that really matter, life altering injuries and fatalities, those will continue to be lost in the world of useless metrics. Most companies, on their journey to zero, will say they've hit a plateau. They have extended periods of time where they have no serious injuries and seemingly out of nowhere, they have a fatality or catastrophic incident. Along with this phenomenon is a set time frame, based on the previous data you can expect a fatality to occur within a certain timeframe. Every 5-7 years, or some innocuous number, that the company believes is the time when we can expect a major event. So how do we advance from here? To start we need to talk about how we got here.

Safety is regulated by law, as a result there are prescriptive programs to get you to compliance. However, we believed there is a land *"Beyond Compliance"* where the promise land of zero injuries exists. This is how we've got here, what got us here must continue to get us where we want to go right? This mindset has created cookie cutter safety management systems that allow us to feel like we're on the right track because our peers are doing it the same way. But ultimately that is just confirmation bias. We are flat earthers at this point because we found someone else that believes what we believe, therefore it must be true. Company X and Company Y are using *Behavior-Based Safety*, observation programs, safety committees, pre-job assessments on *iPads*, etc. The feeling of being the only one is not welcomed in safety. That's where the solution lies, what will fit for us!? And us only! Instead we believe, it must be right if we are all doing the same.

Unfortunately, safety is not something that can be measured and controlled like a bottom line on a balance sheet, the thinking that we can affect that result by doing more stuff is exactly why we keep killing people. Remove Safety from the bottom line thinking and understand it is not something that can be influenced, quarter over quarter and year over year. It's dynamic and looks different from moment to moment. No moment or sequence of events should be looked at as identical. However, with safety management systems today, we have convinced a lot of companies that management can influence safety in the same way we influence metrics on the balance sheet. Here's an example: Let's say we have a company that wants to increase the operating profit from last year. That can be basically done by doing one of two things, increasing the revenue generated while maintaining the current costs it takes to run the business or decreasing the costs that it takes to generate the same revenue. If we do one or the other, we'll make money. We'll be able to see results that can be measured, great! In safety we do the same thing. We apply a safety program, more employees, more something...only to compare two moments in time and decided if we have improved. This is futile because metrics don't matter as much as we think they do, why? Well for one, a dollar is a dollar on a financial statement. A dollar will never change in value. If you move the number of dollars up or down, you can quantify the impact. No two injuries or incidents are the same. However, they all go down as a tick mark on the ledger and get added as an aggregate number for the year. Let's start having tough conversations about what our real goal is, reduce the likelihood of killing our employees. Frankly it's going to be a negotiation, not a conversation because of the sanctity of safety. You're talking about jeopardizing lives, only in the eyes of management. Let's talk how a negotiation will go.

There are two types of negotiations, competitive and cooperative. Competitive negotiations are akin to contract negotiations and even simpler, car buying. Both parties have a goal in mind and are competing to capture as much of the available opportunity as possible. This type of negotiation is typically not conducive for long term relationships because a perceived negative outcome could sour future negotiations. The other type of negotiation, cooperative, is more collaborative and more for long term relationships. The group has a common interest in the outcome, even if their respective objectives are different, and even opposed in some cases. In the end, when both parties depart from the negotiation, a successful outcome is when both parties feel like they have won. That's the outcome that needs to be sought after when confronting management, both parties need to feel like they've won and have more than they started with.

At this point I hope it's been made clear that chasing zero gives us the results we want, but not what we actually need. It also alters our perception about what success really is, it's making sure no one's quality of life is altered, or worse, dies on our watch. Everything else, we can live with. Well, you might say, that's a pretty low bar. Think about it, bones can heal, stiches can be given, but you can't unburn a body, reattach limbs, or bring someone back to life. However, in the safety profession we've been conditioned that zero harm is the goal. We should NEVER see anyone get hurt at work, aka zero injuries. That's just something good to hang your hat on, it's not achievable or reasonable. However, when you look at how many life altering injuries and fatalities that occur every year at work, that's a number worth looking into.

WTF is a Safety Professional?

The true role of the practitioner is a question that has long plagued the safety profession and the industries that it serves. Without a clear and concise definition, the role of the practitioner has been a catch-all for what seems to be nearly anything and everything. Safety has become the company junk drawer; a home for things that can't be placed within other departments. This flawed approach has resulted in ill-feelings all around. Doomed safety professionals face unrealistic expectations; they handle things well outside of their expertise and spend their days putting out fires. Organizations feel perpetually let down by practitioners that fail to meet their definition of what a practitioner should be. Simply put, the current definition of a health and safety professional is not working for practitioners and employers alike. It appears that the profession, both forcefully and sometimes willingly, has taken on too much with too little. This definition, the one where the role of the practitioner has bled into nearly everything, has proven itself to be unsustainable. This mindset that nothing can be left untouched by the practitioner will always be doomed to fail.

Let's start by defining the current view of what a safety professional is, and the role that they often play in organizations. The role of the safety professional can typically be summarized as:

> *An all-knowing guru, a selfless sacrificer, a soothsayer and predictor of accidents, and fixer of company woe.*

But why? How did we end up in this position? How did this become the widely accepted definition of the role? How did this become the norm? The safety practitioner is expected to be the master of nearly everything; they are expected to be everywhere all at once. The practitioner often finds themselves thrust into situations as an "expert," even when they have no expertise in a particular area. They are commonly expected to willingly sacrifice their personal well-being, families, and personal relationships, all in the sacred name of "safety." They are regularly expected to be a lone wolf, handling anything that comes their way, on their own, and without help. If you scour the job boards for more than a moment, you will find a common phrase around safety and health postings: *"Collect and analyze data, interpret information, and predict and prevent incidents."* So, the practitioner is not only expected to be a technical expert in nearly everything, but they are also expected to be a safety prophet and fortune teller. As if martyrdom and faux expertise were not enough, the professional must also issue tarot card readings to the organization in order to predict bad things on the horizon. The practitioner is expected to be everywhere, do everything, and always execute perfectly. Again, how did we get here?

The answer: underlying assumptions within ourselves, our industries, our companies, and the greater safety community about what defines "safety" and the role of the practitioner within that definition. The current description of the safety practitioner is merely a symptom of deeper-rooted beliefs around the general definition of safety. These underlying assumptions influence and shape the outwardly facing definition of the practitioner, how they fit within the safety system, and what falls within their wheelhouse. Let's examine a few of these deeply rooted assumptions:

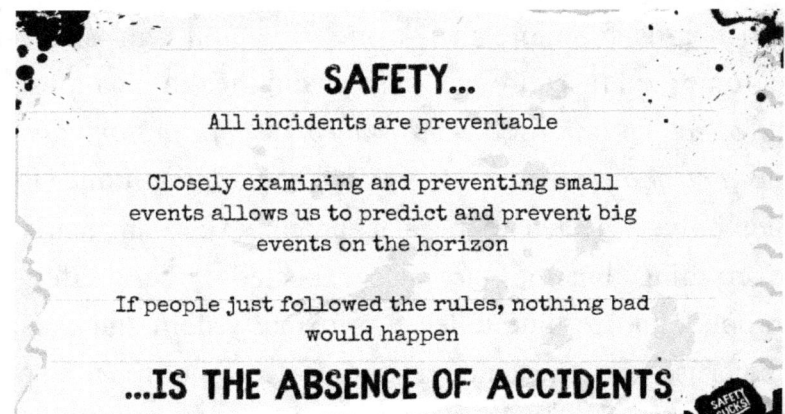

From these deeply rooted underlying assumptions about safety, we craft our definition of "safe." "Safe" is commonly defined as the absence of bad things happening. A lack of accidents equals "safe;" the presence of accidents equals "unsafe." From these assumptions and beliefs grow our organizational values. What do we value in traditional approaches to worker safety? ZERO! Organizations ultimately value outcomes, and safety professionals work to support the values of the organizations that they serve. From our assumptions about safety in general, grow our assumptions about the professional practice of safety:

-We assume that-

Safety is an outcome to be managed
Safety is frequently defined as the absence of negative occurrences. We often seek a desired state of "zero."

-So-

The safety practitioner manages safety
Outcome based safety, along with commonly held beliefs about the role of the practitioner, result in the safety professional managing towards the desired outcome.

-Which results in-

Safety Fix it!
The practitioner's role as a fixer and manager of outcomes is solidified. Nothing can be left untouched by the hands of safety; safety is everywhere and so must be the safety professional.

These assumptions have shaped our modern view of the safety practitioner. Over time, with each passing accident, more and more has been pushed to the practitioner to manage. New responsibilities are often added, and few are ever taken away. The role of the safety professional has grown and grown, reaching bloated extremes. After poor audit findings, observed "Unsafe behaviors" of frontline employees, or general operational surprises, the safety professional is quickly tasked with tending to the unexpected result. Companies have leaned into the notion of *"Safety Fix It!"* The safety professional has become the easy button for many organizations, a singular and highly visible point of both action and blame.

Many safety professionals are quick to jump through the hoop of *"Safety Fix It;"* this is one of the few times that they are truly empowered to affect change upon their organizations. The practitioner's previous pleas for change and betterment are often left ignored, at least those ideas that reach beyond surface-level problems. Then suddenly, after something has gone awry, they are empowered to go forth and "fix" things. Even though this empowerment is only temporary and often insincere, it can be intoxicating to a practitioner that has long dreamed of bringing about change. The organization lays the problem at the practitioners' feet; they have now been granted the authority to mop up and fix the problem. Along with this authority, also comes added responsibility. As the safety professional dutifully begins their task of "*Safety Fix It,*" the organization demands swift and measurable action by the practitioner. The safety professional often finds themselves overwhelmed and unsure; they quickly push out corrective actions to appease the powers that be. But the wrong changes, brought about by the wrong persons, for the wrong reasons, that is only change for the sake of change, rather than change in pursuit of betterment.

We have painted a vivid picture of the current role of the practitioner, how we arrived at these conclusions about their role, and touched on some of the "safety hangover" we face because of it. Now, let's redefine what it means to be a safety professional. Here is a better definition:

> *A communicator, facilitator, team member, and team builder. A curious person with an obsession for learning about work, and an evangelist for organizational betterment.*

- That is a massive shift in views about the role that the practitioner plays. Imagine the difference between these two vastly different jobs, the practice of safety currently v. practicing *safety better*. Much, if not everything, changes. But how do we get there? How can we move beyond our current and deeply ingrained definition of the safety practitioner? It is often observed that a mere change in definition never really leads to meaningful change. Things look different, but much is actually the same under the surface. This happens because, rather than focusing efforts on the source, time is wasted attempting to fix symptoms. An understanding must be maintained that beliefs come from somewhere; aim must be taken at the cause of the beliefs. There must be targeting of the same things that led to the current definition, there must be a change in the assumptions that hide behind the outwardly facing definition of the safety practitioner. Let's explore a path of better assumptions:

-We assume that-

Safety is the presence of defenses
Safety is not defined by a number; safety is the ability to fail, respond, and recover gracefully.

-And-

We build defenses through learning
Learning how normal work occurs is vital to understanding missing, weak, or flawed defenses.

-Which results in-

Learning is the overriding value of the organization
The organization values learning above all else. It is widely understood and accepted that learning is the only tool to create safety.

-So-

The practitioner is viewed as facilitator of learning
As a communicator, facilitator, team member, and team builder, the practitioner supports learning throughout the organization.

This dramatic shift of assumptions would reveal an entirely new set of expectations around the professional practice of safety. The facilitation of learning would now be found at the heart of the job description. Learning, along with operational curiosity, innovation, and the constant pursuit of organizational betterment, would make its way to the forefront. Less important and less impactful tasks such as compliance and paperwork would naturally fall to the back of the line. *"Safety Fix It!"* would fade, and at the very worst be reserved for dire or extreme situations. Even in the event that *"Safety Fix it!"* was utilized, these occurrences would be viewed as learning opportunities to further eliminate *"Safety Fix It!"* from the organization. As learning becomes a value of the organization, the role that the practitioner plays in supporting that value would manifest as an artifact of the underlying assumptions held about safety.

Where to start? We eliminate bad ideas through the continued introduction of better ideas; that simply means that we create betterment through conversation. A real and honest dialogue must be started and maintained around current assumptions and the assumptions that are envisioned to replace them. We cannot just "will" these assumptions into being, conversations must be had, and minds must be changed. Debate must occur and learning must happen. This shift will sometimes move excitingly fast; often it will move painfully slow. It will ebb and flow, and sometimes it will even stall. But focus must be maintained on shifting assumptions through conversation; all change is in fact one large public relations campaign and should be treated as such. By continuing to have these "hard" talks in an upfront and honest way, we drive towards that shift in assumptions and cultivate betterment. Additionally, these conversations will lead to more discussions and dissent around the topic and can result in that much more learning.

Very few, if any organizations, really think about the role that the safety practitioner plays within their companies. The practitioners job description is eerily vague in certain areas, and stunningly detailed in others. Worse yet, it is formed and molded with extraordinarily little thought as to what the safety professional will actually do within the organization. Job descriptions are built in-keeping with the current definition of professional safety; *nothing can be left untouched by the hands of the safety professional.* This results in many safety job descriptions reading as if they are a concoction of every other role within the organization. This lack of true definition results in misalignment. This misalignment, as misalignments around expectations often do, results in disappointment and strife all around.

This particular pain-point can serve as a conversation starter to begin talks around organizational assumptions. This is a prime opportunity for the safety practitioner and the organization to come together in conversation around the things that influence the organization's definition of safety, and the defined role of the practitioner within the organization. Safety practitioners need to begin the conversation of creating an impactful role for themselves within the organizations that they serve. This conversation will typically yield some compromise, but there must be understanding and alignment in order to move forward.

An insightful exercise for an organization can be dissecting the true reasons why they employ safety professionals to begin with. Oddly enough, many cannot deliver solid reasoning for the role. Often one will hear things such as:

"to take care of safety,"
"to lead our safety program,"
"to look for and fix hazards,"
"to prevent accidents,"
"to keep our people safe!"

...and other similarly vague and *"Safety Fit It!"* oriented definitions. All of the above responses can be loosely translated into, *"we are not really sure."* They also precisely align with the views that *safety is an outcome to be managed,* and *the safety practitioner manages those outcomes.* Many organizations have become so focused on the practitioner being everything, that they have no clue as to what the practitioner should be. Focus must be directed back to the better definition of the safety professional to find better reasoning for the role. The organization, along with the safety professional, should be asking one simple, yet challenging question: *how do we get the most impact out of this position?* With a new set of assumptions, a fresh definition of the safety practitioner, and that question in hand, an organization can begin to construct what an impactful safety role looks like within a particular company.

The greater safety community plays a hand in molding the defined role of the practitioner as well. Professional organizations, institutes of higher education, community groups and clubs, and similar, all play a part in the creation and maintenance of assumptions about the professional practice of occupational safety and health. These institutions within the community of practice can either help or hurt in this effort to better define the role of the practitioner. They can perpetuate flawed safety beliefs, or they can drive towards *safety better*. These groups can help lead the evolution of the safety profession, or act as gatekeepers of the status quo. Similar to organizations that employ practitioners, institutions of the greater safety community should be asking themselves about the true role of the practitioner. This is in no way a call for some detailed cookie-cutter definition and list of responsibilities of a safety practitioner. This is not a question of specifics; it is a question of essence. At the core of it all, what does it truly mean to be a safety practitioner?

Now that we have a better description of the practitioner and some thoughts on bringing that new role to life, let's talk about how the safety professional fits into organizational structure. With a better-defined role should also come a clearer reporting structure for the practitioner. This is not meant to be prescriptive as much as it is meant to be informative. There are limitless ways in which you can weave a practitioner or safety department into an organization; these are merely a few examples:

Example I – *Hard-Line Safety with Soft-line Operations:* In this reporting structure, the safety professional reports directly up through several layers of safety overhead. They also maintain partial reporting to several non-departmental leaders such as an operations manager or similar.

Example II – *Straight Line Operations:* The practitioner is a direct report of the leadership team that they serve, typically reporting up through a site or location manager.

Example III – *Support Organization:* The safety practitioner reports up through a safety manager who then reports into a director that oversees multiple support departments such as environmental or similar.

EXAMPLE I
Hard-Line Safety with Soft-line Operations

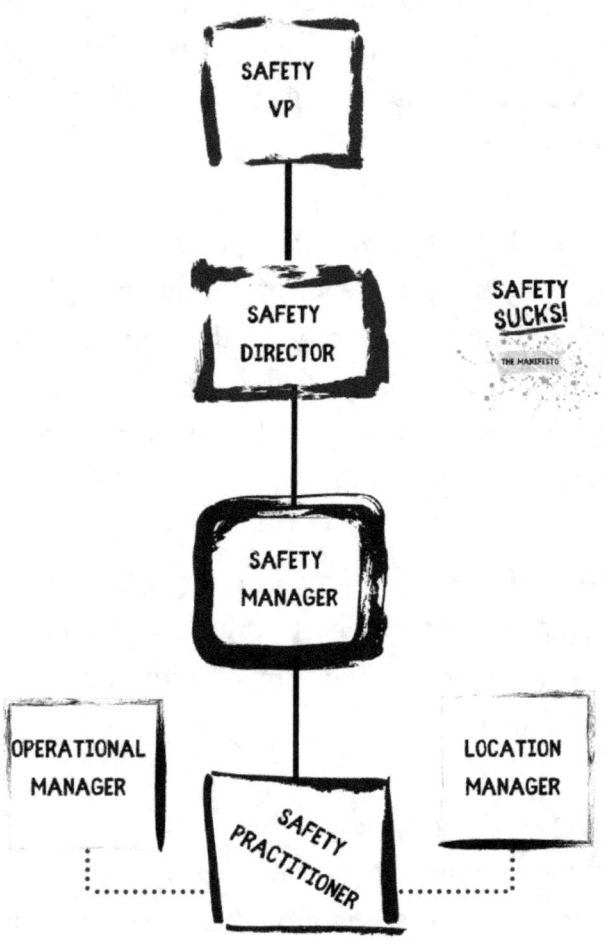

Example I
Hard-Line Safety with Soft-line Operations

Select Pros:

- A large safety support structure for the practitioner
- A direct means of "bubbling-up" issues and ideas up through a safety business unit
- If a healthy relationship exists between the practitioner and "dotted-line" management, the practitioner is positioned nicely to affect positive change

Select Cons:

- "Dotted-line" accountability can often lead to confusion and unclear direction
- Multiple levels of safety can lead to increased safety bureaucracy
- Can cause and "Us v. them" dynamic to develop and result in a power struggle between the safety department and operations

EXAMPLE II
Straight-line Operations

SAFETY SUCKS!
THE MANIFESTO

Example II
Straight Line Operations

Select Pros:

- The practitioner can be viewed as part of the team rather than an outside entity
- If a healthy relationship exists between the practitioner and direct operations manager, the practitioner is positioned nicely to affect positive change
- Simple and can eliminate levels of safety bureaucracy

Select Cons:

- Lack of safety support can often leave the practitioner feeling like a "lone-wolf"
- Safety can become stagnant due to lack of diversity in thought and opinion
- The practitioner has extremely limited authority and lacks the mechanism to gain authority by raising issue through their reporting chain

EXAMPLE III
Support Organization

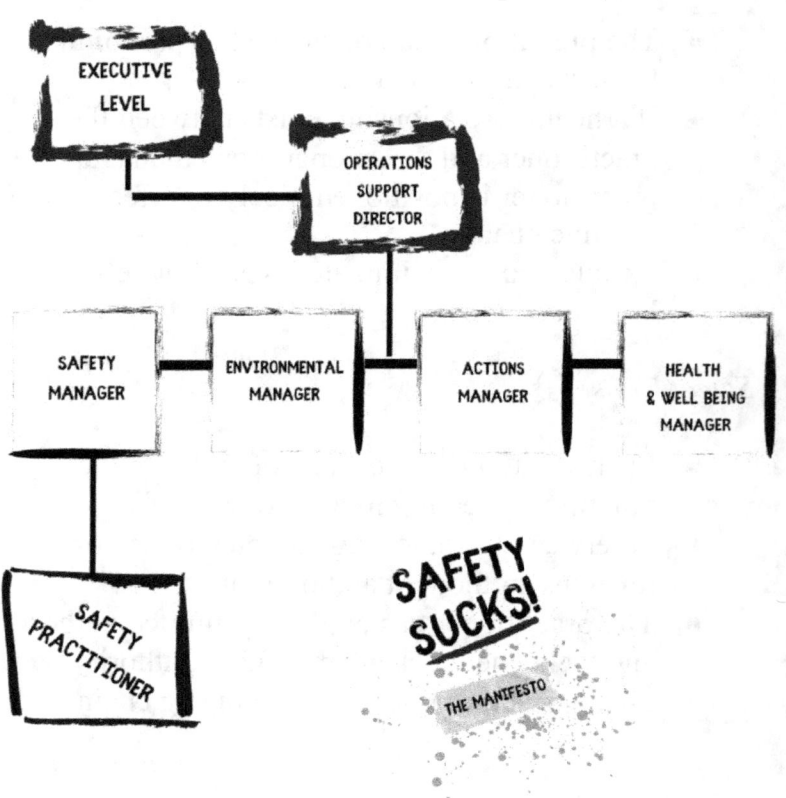

Example III
Support Organization

Select Pros:

- The safety practitioner is often viewed as a consultant rather than a fixer. This can help to minimize *"Safety Fix It!"*
- Can create an environment conducive to the formation of strong safety teams
- The practitioner can easily reach out to other support functions for support

Select Cons:

- Directors and managers within this hierarchy often lack knowledge on the practice of safety
- Safety could find themselves delegated tasks that leadership does not know where to place elsewhere – Can further promote the junk-drawer effect
- Support departments are often business unit specific. This leads to siloing of various safety departments in a single organization.

However you choose to approach reporting, there is an innate need for a clear and agreed upon reporting structure for the safety function. The murky and ever-changing reporting schemes of most organizations have only led to headache for all parties involved. The safety practitioner is often tucked within the organization with little to no thought as to where they are placed. This results in multiple reporting points for a single practitioner and power struggles between those various points. Whereas most employees have a single direct manager, the safety practitioner has many. Confusion abounds and the practitioner and employer are both left unsure of who reports to who. There are a few key points to consider when placing the role of safety within an organization's hierarchy:

- Focus on keeping the role impactful.
- Simple and clean beats intricate, every day; keep things easy to understand.
- Avoid too many "dotted-line" relationships and multi-point reporting schemes. Doing so will help to avoid confusion and ambiguity.
- Consider the practitioners ability to challenge authority and speak truth to power before settling on a structure.

Unfortunately, there is not some "one-size-fits-all" reporting scheme. Providing a rigid and prescriptive plan would only result in further grief for organizations and practitioners alike. With a better definition of safety, a better idea of the role of the practitioner, and some deep and meaningful conversations between the organization and the safety practitioner, a bespoke and workable reporting structure can be developed. Organizations, along with their safety personnel, can micro-experiment and "try-on" various styles of reporting to see what works best for their particular organization. In any case, as with defined role and expectations, reporting hierarchy must be thoughtfully designed and agreed upon. Doing so will only help the practitioner and the organization by reducing confusion, further eliminating ambiguity, and ensuring the practitioner is in the best position possible to have a positive impact on the organization.

The practice of professional safety has long been affected by an identity crisis; a crisis that is in dire need of resolve. Flawed underlying assumptions have led to poor definitions of the true role of the practitioner. These definitions, those that have resulted in the role of safety becoming a company junk drawer, must be evolved towards practicing *safety better*. This is not a question of specifics; it is a question of essence. At the core of it all, what does it truly mean to be a safety practitioner? It is a matter of impact; the current accepted role of the safety practitioner limits the true positive impact they can affect upon an organization. By shifting the role of the practitioner from fixer to facilitator, and from fortuneteller to curious learner, the impact of the role is strengthened. This evolution occurs through targeting and changing the underlying assumptions that have led to the current definitions of safety and the safety practitioner. The formation of better assumptions ultimately leads to a better and more impactful role for the safety practitioner.

Losing my Religion

Safety is held back from progressing because certain slogans seem indelible in our industries. They seemingly demonstrate a level of care by the organization, so they are damn near written in permanent marker, because despite the sign's faded appearance, we can't ever take these down. Taking down a safety sign might show that we don't care! Here are some of the classics:

"Everyone goes home the same way they came in."

"Stay alert, don't get hurt!"

"YOU are responsible for your safety!"

"Nobody gets hurt today!"

"If you see something, say something!"

"Zero injuries is our goal!"

And my personal favorite...

"Safety is no accident!"

These slogans don't mean a whole lot when they aren't plastered at the entrance of a jobsite with the same tone as a similar sign used to deter unwanted behavior, "No Trespassing." Now you may say to yourself, these signs are proof that an organization cares. If they didn't care, wouldn't they have nothing up at all? Can a simple slogan and sign really set back an organization? Yes! Each one of these phrases implies that we care as an organization, but it simultaneously reveals a thought process that is rooted in traditional safety. A thought process that is founded in the idea that employee actions and behavior are directly responsible for outcomes, also known as, "We care about you not doing something stupid, like getting yourself hurt."

Deliberate action needs to be taken everyday by YOU to ensure that YOU don't get hurt. We see this implied in these safety slogans and signs, "YOU need to look out for your coworkers...If YOU see something, say something." Which directly correlates to, if something happens to you or around you, you are at fault because of proximity!? Typically, an organization will pair these slogans with the old poem turned pull on your heart-strings video, "*I Chose to Look the Other Way.*" Which really hammers home the point, safety is

YOUR responsibility. If you complete a job "Safely," that's because it was due to purposeful planning and flawless execution! Also, that logic implies that a lack of safety is something to be punished because the opposite case is achievable. Unfortunately, at this point we are penalizing employees for not being able to see something they have grown accustomed to working with or for not recognizing it as a hazard. But we have the audacity as a viewer, after an incident, who has all the facts gathered and the ability to have time and a team to evaluate a situation, to blame and punish. Not only do we have time, resources, and benefit of the doubt with our assessment, we begin to develop an arrogance about our perspective and ask questions like, "How did you not see that hazard?" Which leads to a simple solution like, putting up signs to promote the desired behavior.

Management feels the need to advertise these slogans to make it known that they care about you and your wellbeing. They care so much that it needs to be on posters, hardhat stickers, orientation pamphlets, signs welcoming you as you enter the facility. Essentially saying, we mean business and we are in the business of keeping you safe! They demonstrate seriousness and commitment to safety. Management can rest easy knowing they did what they could to provide their employees with the right tools to succeed. In the end, all this does is make it look like something was achieved, the physical act of putting up a sign or adopting a slogan fills a void. Where there was a blank wall, now lies a mantra for us to live by, mission accomplished. We are now safer as an organization. Unfortunately, that's about all these slogans do... give the appearance of caring. These phrases give a false sense of a moral safety compass and dictate logic that is misleading about how to protect yourself in the workplace.

So now that we know that signs and slogans are *Fool's Gold*, easy fix. Get rid of them! However therein lies the problem, graduating from slogan-based safety to a more advanced approach isn't easy. Have you ever asked a member of management to take those signs down? By adopting the slogan or erecting the sign, we have opened *Pandora's box*. Management response to removing these signs may be along the lines of: *"NO WAY! Are you crazy!?"* *"Employees won't know we care about safety if we take them down."* *"Not only will we NOT take them down, how dare you imply that we would consider your request!? Blasphemy!"*

This exemplifies the conundrum we face at our workplaces. Damned if you do, damned if you don't. But also, this highlights the fact that doing something feels better than doing nothing. Organizations need to view safety more as a science and less like a religion. What do we mean by that? I'm glad you asked!

Religion:

Religions set forth a foundation of morals and beliefs to live by, often passed down from previous generations and believed due to general consensus among a community. Regardless of the religion, there are rules and recommendations to help one live a life that's in accordance with the overall goals of that religion. The rules are typically hard lines that are drawn to set boundaries, to determine what is right and wrong, these guardrails act as ways to keep you on the path of the righteous.

So, at this point you're probably wondering, what the hell does this have to do with Safety!? The notion that safety is somehow analogous to religion seems like a stretch, but when we examine closer, this isn't farfetched…

Safety as a religion:

I believe in safety because it gives me a foundation to know what is right and wrong. It helps me determine what can protect me in this dangerous world that we work in! Its mere presence gives me comfort in knowing that I'm taken care of today and every day. I have faith that I can prevent all injuries in the workplace, I am in control of my destiny. I need to believe that I can't get hurt and take every step to make that happen, it will be so. Every incident is preventable, and every injury is avoidable.

Let's add another layer to this, I'm not picking specifically on Christianity, but the idea that ten commandments can govern a life is fun to poke fun at as it is so close to our world of safety that it needs to be addressed. Let's use the *Ten Commandments* as an example. This is a great way to establish guidelines and display the most sacred rules that shouldn't be broken, no matter what! While in safety we don't usually have ten items, we'll have four or five. Instead of calling them commandments we'll call them something like "*Rules to live by*" or "*Golden Rules*," because if we decree that these rules are never to be broken, we will see desired outcomes. In this case the desired outcome is not seeing anyone get killed at work. Not only will we never kill anyone, but we will also fire anyone who breaks these rules! If you're not willing to follow our commandments, err rules, you clearly can't be a part of this religion/company. You're not safe enough to belong.

I will acknowledge that having *rules to live by* at face value sounds like a great idea. If you never want to see an employee killed by falling from an elevated work surface, make a rule so it never happens. Let's play this out however, further proving why slogan-based safety and *rules to live by* is a hindrance. So, your company wants to implement a 100% tie off policy, in addition this policy has a zero tolerance for non-compliance clause. This basically means that anyone caught not tying off will be fired immediately. When this is presented to the leaders of your organization, this demonstrates caring and reinforces the idea that we are holding strong to what we believe in, caring. We are going to care so much that we will fire you from our company if you don't follow our rules because we'd rather see you unemployed than dead on our watch, which sounds admirable at face value. However, while we can take solace in the idea that since we fired someone for not following our rules, that doesn't reveal the truth. The truth is, you just caught someone doing it, that doesn't mean these *rules to live by* are helpful. It just means that your sacred rules, commandments, are being broken more often than you think. We can lean on the proverb, *"One often meets their destiny on the road to avoid it."* Bottomline, these rules to live by do more harm than good.

Why are these sacred rules in place? There is an assumption that if we write it down, hold people accountable, print them on laminated cards, make neat posters and plaster them all over our facility… we'll never kill anyone. A generic list of rules to live by:

>  **IMAGINARY XYZ CO.**
> Safety is Number 1!
>
> Thou shalt always do a live-dead-live
>
> Thou shalt always tie off when working above 4 feet
>
> Thou shalt never work within the minimum approach distance of energized electricity
>
> Thou shalt always exercise stop work authority!
>
> Thou shalt never work without a LOTO
>
> Thou shalt always wear PPE...always!
>
> Thou shalt never drive unsafe or distracted

(Seven is a lucky number, so let's just end there.)

The hardest lines you draw in your organization will yield two guarantees, you'll create a perception of compliance and you'll never have learning events. By assigning so much weight to these ideas, you'll essentially tell your employees, "I'd better not catch you breaking these rules, but also I shouldn't even catch you thinking about breaking these rules." This hypervigilance will cause employees to look at management and supervision like the judge, jury, and executioner. So, if they have to break one of these said sacred rules, there's no way in hell that they'll mention it to anyone. Even if they felt it was needed to complete their job, they'll make sure their boss doesn't know!

Organizations need to stop drawing hard lines in the sand, these pseudo commandments kill people. Having a commandment that mandates *Thou shalt never work within the minimum approach distance of energized electrical sources* seems like a great thing to write down on paper. However, that's assuming that the employees who do the work don't already know that.

Organization: *"Hey journeyman Lineman, did you know that if you get too close to that, it might kill you? We're gonna create a rule to live by so you know not to do that."*

I don't know exactly what the response would be, but it would surely result in something like...

Lineman: *"Yeah, no shit Sherlock... Thank you Captain Obvious... What else are you gonna tell me, don't stare at the fucking sun?"*

THESE EMPLOYEES KNOW THEIR WORK IS DANGEROUS! If you tell someone, "I'd better not hear about you breaking a rule," You know what's going to happen… you won't hear about it! This is the problem when safety is treated like a religion. We expect compliance to maintain order. If those sacred rules are in place, you'll never hear about opportunities to make things better. Eventually you'll have a fatality on your site because someone worked within that minimum approach distance and was electrocuted.

Let's walk through how someone could possibly die by doing something you explicitly told them not to do… I mean you put up the rules for everyone to see! As you do the investigation, you'll come to find that on more than one occasion that employee would work within the boundaries. This most likely wasn't their first time. Also, you may even find out that the entire crew does the same thing. Why aren't our sacred rules working? We made it clear they need to never do these things! That same journeyman lineman won't tell you they've worked within the minimum approach distance in the past because precedent will have been set. They don't have a guarantee that the organization will be understanding with their confession that they had to work within the minimum approach distance. Why potentially put your job on the line when you know for sure that if you just don't say anything, there's no way you'll be breaking the rules?

Let's put some context around this, during the investigation you find out the lineman had two choices: 1. *Say something and stop the job,* or 2. *Don't say anything and find a workaround.* We're obviously investigating a fatality so let's look at how it happened. *Option 2.* The crew assumed the overhead line job could be done by using the same exact isolation pattern that worked in the past. However, due to a new technological addition within the system, a permissive now prohibits that section of line to be dead, when using the same process as before. This improvement is to ensure we have redundancy and improve reliability. In actuality, an unintended consequence now exists. This new technology now complicates isolation, and requires the team to isolate the entire system, not just the section they did in the past. Let's play out the scenario with the lineman speaking up and saying something.

Option 1. The lineman tells their boss and is immediately scolded for not having an electrical clearance in place, prior to starting the job. They assumed the job could be done by closing a series of closers and relays that worked in the past. However, due to the new technological addition, a permissive now prohibits that section of line to be dead without isolating the entire system. The crew didn't know this, so they had to call their boss and break the bad news. Not only will this job not be done in the time they allocated, but it's also going to require a substantial scope change and larger isolation plan that will impact other project timelines. In this scenario, this crew got scolded for not keeping the schedule, not knowing beforehand that they needed a larger clearance, and getting their boss in trouble. Now the whole job needs to be re-evaluated. The boss finally relays the message to the affected groups, we have a lot of work to do, they shut the job down, wait for phone calls, put the necessary clearances in place, get the right tags hung, and verify a dead electrical state before work can begin. The electrical energy has been isolated and brought down to a level that's safe enough to work within the minimum approach distance.

Flash back to that linemen deciding. Let's go with *option 2*. This person is aware of the implications of *option 1*. Yes, they can tell their boss, but knowing the implications of having to reveal they missed a step, they decide against it. If they say something to the foreman, for sure management will flip their lid! So, what occurs is the lineman goes within the minimum approach distance for a second, "What's the harm? I know it's safe enough. Save some time and avoid my crew getting chewed out." But since this is a *rule to live by*, you'll never hear a peep about it. You'll never learn that the crew fears stopping the job for a clearance because of improper planning, poor scope, or fear of shutting down other jobs as a result. So, the minimum approach distance line gets stepped over once, and what happened? Nothing! So, since it happened once with no consequence, one time becomes two, and then three, and eventually it becomes acceptable to do so with the understanding that you tell no one. Why does this continue? Don't they know it's dangerous? Well, they have no incentive to stop this practice. They're getting praise from your organization for executing on time and if they say something, they may lose their job. Therein lies the problem with the slogan-based governance, safety as a religion.

Safety as a religion has no spectrum of compliance. There is no room for almost following the rules or making the rules work for the job. It's absolute. There is no context. Sin is sin. Commandments of safety, thou shalt follow the rules, no matter what! If you are caught glancing at your phone while driving, you're just as guilty as someone working within minimum approach distances... you shouldn't be working here. So why would the Lineman in the story ever mention that they've had to work within the minimum approach distance AND why would they ever mention that they have a history of doing it?

Why go through all this trouble to understand that safety is treated like a religion? We must first understand the issue before we solve the problem, the issue is we need to view it more like a science. A scientific approach attempts to view situations on a case-by-case basis. Each instance is unique and a set of assumptions that were present from the previous similar situation may or may not apply to this situation. Be willing to admit that the rules that govern the work are wrong. Blasphemous I know. Understand that the employees will appreciate you understanding their struggles and will be more likely to bring you more problems that you'll never unearth on your own.

Looking at safety as a science is necessary for understanding the layers of truth that exist but aren't seen by those who don't do the work. Most organizations don't know how to apply science to safety because it is seen as a purely qualitative exercise. It's a "Feeling." A general understanding that is assumed but is ultimately relative. What is a hazard to one person, is everyday life to someone else. The word itself, "Safe" is relative from person to person. Science will explicitly define what safe means. Obviously, there is no single definition for anything, as we work across industries, cultures, and regulations. But agreeing on a definition is critical to communicating.

Safety as a Science:

Workplaces are inherently dangerous and should be recognized as such. Assumptions will only be made once evidence-based theory has been applied to an assessment of the issue. The workplace occupies an environment riddled with opportunities for failure, so a capacity factor for said failure must be built into the work. The only assumption that should be constant is that events can't be anticipated. Workplaces should do their best to anticipate the most serious hazards and create capacity for failure, but also should not expect to guard from everything because it's impossible to know everything."

Safety as a science doesn't just apply to incident investigation. This also applies to management of our programs we implement and maintain, data analytics, along with the language we use. Let's begin with incident investigation.

A manufacturing facility has a big leadership team meeting upcoming. The plant manager is preparing to leave for headquarters, which is located two hours away by car. He plans to leave at 9AM, get there at 11AM, just in time for a quick bite to eat before the meeting begins at 1PM. Once the meeting is over at 3PM he plans to drive home to finish the day. While he is en route to the meeting, his vehicle is rear ended while he is sitting at a red light. The driver of the vehicle that struck him was unaware that the light had turned red, he was texting and driving while travelling 40 miles per hour. The plant manager sustained injuries to his back, neck, and has suffered from PTSD ever since the incident. Now I ask you, how was that incident preventable? You have a plant manager out on long term disability due to an event that occurred while he was traveling from one work location to another. There is no way he could have anticipated or prevented this incident. Rear end collisions roughly occur 1.7 million times each year in the U.S., trust me I couldn't believe the number either. Simply from an odds standpoint, due to this many "shots on goal," it is bound to happen to someone while they are working and likely that it'll eventually result in a serious injury. You know what companies typically do with that injury? It is filed away as an act of God, and usually a bulletin about defensive driving will be sent out to the employees. Just to make sure that we are covering our bases!

This seems to be one of the few times we actually apply science to the workplace, why? The reason is, we all drive. We all know it's dangerous. We all understand that we have the potential to be rear ended at a red light. If this happens 1.7 million times a year, it makes sense that eventually we or someone we know will be affected by this event. So why don't we ask for new rules for driving to ensure no one will ever get rear ended again? Easy, because any rule we implement will impact the time it takes to get to our destination. This same type of event occurs every day at the workplace, where the worker does not have control of the environment around them. However, because we have put a boundary around the workplace and called it our property, for some reason we assume that we can control everything. It would be silly for us to assume we can control all hazards and potential hazards around us.

Next time we evaluate an incident we need to look at it from the standpoint of, what were the capacities for failure and how effective were they? Let's go back to the rear end example. We have airbags and seatbelts. Typically, these reduce the overall impact of the incident. Most times, when someone is rear ended at a red light, these can prevent death. The seatbelts and airbags often reduce overall injuries to muscle strain, and maybe a black eye or two. While those are still not desirable, the capacity for failure was effective! The worst outcome in that event is death, no seatbelt, or airbags. But by simply adding those in, you can reduce that consequence to a soft tissue injury and a few bruises? Amazing! We need to begin adding some rigor to how we approach incident investigation and betterment at our organizations. Organizations suck at implementing and maintaining safety programs. A scientific approach is drastically needed when considering any program because currently we have shotgun solutions. A program that "Feels like the right thing to do" will always supersede taking the time to find the right solution(s). Great, we have an idea! It must be quickly rolled out to the crews! Quick, send it out. Hurry! We're in danger of being late to the party of being safe!

Okay so let's go with an example to drive the point home. Let's say we have an electrical line inspector, tasked with inspecting power lines for damage, wear, and significant issues. From time to time, they need to stop and get out of their vehicle to walk along a barbed wire fence and perform minor repairs in the field, get closer inspection of an issue, or get exact longitude and latitude for work orders. In the process, an injury occurs due to an employee cutting their palm on the top of an old, jagged T-post while trying to climb over a right-of-way barb wire fence. This happens more often than you might think. Typically, the openings in the barbed wire fence for vehicles and people are miles apart from each other. If an employee forgets a tool, material, or sees an opportunity to fix something on the other side of where they are standing, the easy solution is to step over the top barbed wire like you're stepping over a pet or child gate in your house. Yeah, you could go all the way to your vehicle, walk to the nearest opening, or leave the issue to the next person. However, climbing over or crawling under is common practice, everyone does it. So, this employee climbs over the fence, all while balancing themselves using the T-post that the wire is secured to. While swinging one leg over, their grip slips with all of their weight on their palm and boom, the post rips through their standard leather work glove and stitches are needed to close a gash across their palm.

This is an example of an employee that does something because it makes sense. They probably get praised for being quick, efficient, and being a "problem solver." They work alone, no need to ask for help during a simple inspection route. The company can't give them another employee for inspection, that's a solo job. So, they decide to mandate thicker gloves, *Kevlar* to be exact. If the employee had only been wearing *Kevlar* gloves, it wouldn't have torn as they went over the top of the fence! More importantly, we wouldn't have had the injury! Push it out, mandate *Kevla*r gloves for everyone so we'll never have this happen again. Great idea team!

However, the employees resent the requirement to wear the new glove. The glove that you gave to them is thick, it makes for sweaty hands, feels like a cast, and the employees hate it. But the organization can sleep better at night knowing a glove won't be cut through on their watch! We need to experiment with solutions before we shotgun our ideas to the field. First ask, why do employees feel they need to crawl over the top barb to scale a fence? Is there a better way to climb a fence if it truly can't be avoided? Has someone made a product to help with this issue? Has this issue been brought up in the past? What was done about those concerns? Instead, we do a 100% rollout overnight and affect the entire organization. Not only will we shotgun the solution, but we'll also never take the feedback and pull back the glove mandate. We never get to a point where we can admit that a "solution" was stupid or augment the mandate to fit a more effective solution. That blood from the cut palm needs to be memorialized! This is silly for multiple reasons, but strictly from a business perspective, in any other industry you'd be setting yourself up to fail if you performed business in this manner. For a solution that can combat this, we need to look at who constantly changes and does it well, tech.

In the field of data science, *Netflix, YouTube,* and *Amazon* are king. Why? These organizations have understood that they need to constantly monitor the user experience and curate their suggestions to keep you on their site. Screen time yields opportunities for ads and for you to be more attached to their product. The design of a *Netflix* dashboard is crafted using customer feedback. The recommendations for movies, TV shows, and genres use machine learning to gain a better understanding of what types of film and tv you are more likely to watch. This makes you spend more time looking at *Netflix* and less time searching for movies elsewhere. Why is this important?

In data science there is a *70/30 rule* when it comes to experimentation. A test and a training group are identified when measuring the effectiveness of a change. Let's say *Netflix* is trying to determine a way to drive more traffic to their *top 10 list* that's suggested to users. They would like to push *Netflix* produced content onto the users but if they don't use the list, it won't be an effective way to leverage this list. The current model is, showing a *top 10 list* of all movies and television being watched on *Netflix*. The proposed improvement is to change the *top 10 list* to only show the top movies and television shows that are within the genres that a user is watching. One argument for the old model is that people want to see what's most popular, regardless of genre. The argument for the new model is that users want to see what's popular within the genres that they are paying attention to the most. So, to test this, *Netflix* will implement the change of showing the new top 10 to 70% of all their customers. The remaining 30% will see no change and be left with the same *top 10 list*. The measurement is which *top 10 list* will yield more engagement from users. The time interval for this type of experiment, is typically six months to measure. Along with the time, we'll need to establish KPIs (key performance indicators) to measure performance. Once the six months are over, *Netflix* will look at the results and decide which top 10 list drives users to select the desired content from the list. The better outcome will drive *Netflix* to determine if the *top 10 list* stays the same or changes. Regardless of the outcome, another experiment will be happening soon, because they always want to make things better for the end-user. This is a simple example of why companies like *Netflix, Amazon, YouTube,* and *Pandora* are successful. They experiment to see what works and evaluate for effectiveness. Not only do they get it right based on what the data tells them, they continue to improve what they have already improved upon. Dial, manipulate, modify, and augment your solutions to fit your model, in this case the model is your work site. If our organizations ran *Netflix, Amazon,* or *Pandora* the same way we run safety, we'd get a complaint that the site is down and as a result we'd instantly triple our server size to ensure we never go down again. Or, if we got too many "thumbs down" on recommended content, we'd shut down our machine learning engine and go back to allowing customers to only choose their own content, thereby losing the ability to push new content to users. The point is, we suck at making decisions.

Safety needs to be a blend of qualitative and quantitative data. It's okay to have solutions that span across the company, but we can't assume results before we ever roll out the program. How many times have we seen a mandatory policy rolled out due to a slip/trip, muscle strain, or allergic reaction? We need to at least employ experimentation beforehand. It goes without saying, people are not data to be manipulated, crafted, and refined to fit our needs. Also, the employees all talk, so if you have a small enough organization, it's difficult to have a target and a test when the employees can talk to each other and potentially skew your experiment. So, maintaining data integrity is one of the most important parts when thinking about rolling this experimentation process out at your work site. Ensure, as best as possible, that the 70% can't influence the 30%, potentially by segregating the program by location/interaction. The point is, we need to measure effectiveness and find ways to ensure we are measuring just that. As for actual indicators of success, well that could be anything as long as it is consistent: customer/employee satisfaction surveys, randomized feedback interviews, or simply general employee engagement. Do they seem happier with what was proposed? Great, glad it worked out for them! Or do they seem miserable and are simply complying with the change? Well, that's terrible, we need to course correct ASAP!

Let's examine real-world application, let's revisit the electrical line inspector. The suggestion to mandate gloves isn't bad, it's an idea that unfortunately wasn't vetted or tested. Ultimately, the idea came from a place of caring and a desire to avoid repeat occurrences. Where the organization went wrong was mandating a new minimum for everyone, even employees who don't have the same hazards as the injured employee. Let's apply the *70/30 rule* employed by data scientists to this particular situation. So, 70% of the workforce gets the new glove policy, *Kevlar* gloves. In this scenario you're also going to choose a time interval to measure the two sample groups of the population, let's give it six months. Throughout this six-month period, you will document observations of the workforce. What is their general mood? Are they more or less engaged? What are the unintended consequences? What do they suggest could be done to improve the glove mandate? As for the 30%, ask them the same questions. I'm sure in no time, now that we know what the actual reception is, the 70% will express their dislike for the glove mandate, how it applies to only certain circumstances and shouldn't be a mandate for the entire workforce.

By implementing the *70/30 rule* with your experiment, you will also give yourself the opportunity to have an out. Pulling back the program and reassessing will take swallowing of some pride, but it would be the right thing to do. Especially if the data shows that the 30% are more engaged because they saw an injury to an employee, and no one freaked out and did something crazy like mandate a new PPE policy.

Whereas the 70% are jaded because reporting an injury resulted in making their lives miserable and their jobs more difficult to perform. See! You'd be able to have that on paper, juxtaposed the outcomes and you'll immediately see that the glove mandate needs eliminating or serious modifications.

Immediately I know there will be an aversion to this experimentation process because, *"We might be leaving 30% of our employees liable to injury by withholding the solution from them!"* Granted, I can't refute that claim. But back to the *Netflix, Amazon, YouTube* world of data science, they write off that six-month period of experimentation as a business loss for the test group, the 30%. Before they begin experimenting, they know that if the 70% works and they make money off the changes, the 30% is a loss because it never had an opportunity to make the extra money. So, any experiment where process improvement is the desired outcome, your test group will tell you that you're losing money and opportunity with your 30% or your 70%. In this case, we're not talking about money, we're talking about human lives and livelihood. However, in our current work practices, we learn either 100% that our process works or 100% that they don't work. The problem with our current model is, we have no idea which is which. We could have really solved a problem that has a high likelihood of killing someone at work, or we masked the problem even further. While a shotgun method makes us feel better, it doesn't do anything to tell us if we are on the right track. So, I will go back to the *70/30 rule*. 70/30 is what the world of data science employs. That's not saying that safety can't use 80/20 or 90/10, although I'd be wary of overfitting your data. Which basically means you're using a small enough sample size that you could have chosen "that" 10 percent of your population that just does everything well, no matter what you put in front of them. It might give you a false positive that your current/test process is more effective than your target 90%. This is going to take some experimentation on its own to dial in the training/test ratio, the duration of the experiment, data collection method, evaluation of the data, and data analysis. However, regardless of the variables selected, as long as they are consistent it will yield progress. Even if it is addition by subtraction, of programs, rules, mandates, etc.

Bottomline, safety isn't just a quantitative vs. qualitative profession, it can't be whittled down to a simple idea like Safety as a Religion vs. Safety as a

Science. This juxtaposition is presented in this chapter in hopes to illustrate that there's good and bad in everything. The qualitative and quantitative each have flaws and aren't effective without the other. Our job is to determine what blend of the two is going to be effective for our leadership, culture, and workforce. Exercising that discretion is what makes our job so difficult. We need to be observers of people, management consultants, psychologists, data scientists, research associates, and project managers. The solution is relative, yes, but that doesn't mean there is only one answer to our problems. That is the beauty of safety, not only do we not need to have all the answers, but they can come from anywhere. The truth is Safety professionals have a nearly impossible task. They are called upon to answer questions that aren't fully understood by those asking seeking truth. Within that room of stakeholders, a safety professional must figure out how to ensure that we don't hurt people while not screwing up relationships and confidence of the people who actually do the work. The answer is muddy, but a combination of experimentation, ditching slogans, letting go of old programs, and never giving *rules to live by* is a heck of a place to start.

If we get to the point where we are innovating, experimenting solutions, getting rid of blanket solutions, and giving individual crews the autonomy to govern themselves; that means we are inherently recognizing that our people and workforce have identities. They in fact aren't a computer program that you can enter a code/rule that applies to all and is applied equally. Employees can see when you are doing things in their best interest, or that include their perspective. Ultimately, treating safety as a science is just another way to do something we all do as safety professionals, listen to the employees who do the work.

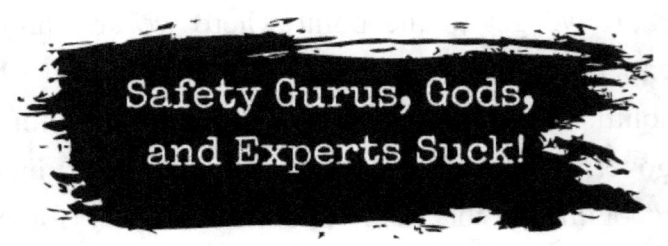

Safety Gurus, Gods, and Experts Suck!

Many unreasonable labels and expectations are often placed upon the professional safety practitioner. From the expectation that they be the selfless sacrificer, the one who alleviates others suffering in exchange for their own, to the demand that they peer into the future as safety fortunetellers. They are to be a constant overseer, one that is ever present and always there in the nick of time to see and stop bad things from happening, and they are to be the judge, jury, and executioner when they discover someone violating our most sacred *Rules to Live By*. Even more troubling and dangerous, and even more common, is the fact that they are expected to be safety gurus, shaman, priests, and experts. Organizations, employees, industries, and even other safety practitioners often demand that the safety professional be the "great knower of things" for safety and much, much more.

We regularly build our safety management systems around this notion, often inserting the safety professional as both the final authority, and the final approver, for most things that occur in the workplace. I'm almost certain that you experience or have experienced this firsthand. We demand that employees not perform certain tasks without the all-knowing input of the practitioner, as we believe that the safety gurus' eyes can see hazards that some lowly workers cannot. We refuse to allow employees to complete the most sacred of tasks such as checking a confined space with a multi-gas monitor, inspecting certain pieces of equipment, examining particular tasks for hazards, or even crafting safety messages, because only the safety expert is suitable to perform such vital things.

Many organizations do not seek to employee safety professionals; gurus and experts are much more desirable. The desire to have a "great knower of things," one that the organization can turn to for all the answers about the safety universe, demonstrates our misunderstanding of how work actually happens and our desire for "easy button" safety. Unfortunately, when the safety deity ultimately stumbles and does not know all, or provides the wrong answer, the organization turns against their safety god by promptly and publicly lashing, beating, and crucifying them. The industrial lore of the failed safety messiah then moves quickly, serving as a cautionary tale for other safety practitioners to never slip in their divinity.

This idea, the one that says safety professionals must know all, see all, and be the ultimate subject matter experts of all things that land in the safety junk drawer, is inherently flawed and results in harm. Let's start by defining some of what we're talking about:

Expert
One with the special skill or knowledge representing mastery of a particular subject
Merriam-Webster

Guru
A teacher and especially intellectual guide in matters of fundamental concern
Merriam-Webster

Deity
Divine status, quality, or nature
Merriam-Webster

The safety practitioner typically finds their role being described with some combination of these definitions. This combo platter of descriptors represents the ultimate desires of the organizations that the practitioner serves, and their beliefs around the value that the safety professional should bring to the table. In addition to the current definition of the safety professional, one that we have already described, the practitioner is often viewed by organizations as:

Safety Guru

A divine and infallible intellectual guide in all matters that relate to our most sacred belief, safety. A master and teacher, one that possesses unique ability beyond that of mere mortals, in all areas of safety and health.

The professional safety practitioner finds themselves thrust into this Christlike status because organizations, industries, and often safety professionals themselves, regularly see this as the most valuable use of the position. Along with this Christlike definition comes Christlike treatment, the safety professional is frequently required to "die on their cross" for the sins and incidents of the organization. We have managed to meander our way back to a bit of the current defined role of the safety practitioner, the *selfless sacrificer.*

How did we end up here? Let's first examine the perspective of organizations that employ safety practitioners. As we have already mentioned, ease is definitely a factor. In a complex, complicated, and chaotic world, companies crave easy. Anything that promises easy solutions, even those things that are obviously flawed, are regularly embraced. Safety is a high priority area for most organizations, but an area in which they have extraordinarily little true understanding. The misunderstanding of how safety manifests in their workplaces, coupled with their extreme motivation to improve safety performance, leads many good organizations astray. The desire for simple solutions to complex problems leads them down what appears to be an easy and morally sound path. Organizations regularly seek to solve complex problems through simplification. This phenomenon is easy to observe, it is often on full display in various areas of the business. But in no area is it more blatant than in safety and health: Did someone do something that we didn't like? We just need to write a rule! Did someone cut their hand? We just need a new glove! Did someone get killed? Well, they should have just cared more about not dying at work! Wait, people aren't wearing fall arrest protection? Easy-peasy! Make it a *Golden Rule* and fire them if you catch them not wearing it! So, we have too many recordable events, eh? Easy fix! Tell people to not have that many, and if that doesn't work, demand that the safety gods case manage them out of being recordable! There, we just solved safety! Yes, the book is over, we're done here.

As you look back at the "easy button" approaches that many organizations choose to pursue, it's just as easy to chuckle at the absurdity of it all. But I'll ask you, is it untrue? It's hard to laugh when it is the world that many organizations create. Easy has become king in a world of complexity and chaos; easy is comforting and it feels doable. This perceived ease pushes organizations towards the safety elite, gurus, shaman, and gods. Answers poured down from above by safety experts are simple, easy, and require no added work by leaders or others within the organization. A trip to the holy cathedral of safety, the safety office, is much easier than seeking out and finding answers through learning from those who actually do what we seek to understand. Not sure what to do? Just ask safety! Not sure what your employees are up to? Just ask safety! Are those pesky incident rates cramping your style? Just ask safety! Easy-peasy-lemon-squeezy!

To take this one step further down the path of exploring these "easy button" approaches to safety and the views of the safety practitioner, let's talk about blame. Having a safety god seems great to organizations until that god gives an answer that they do not like, or the safety god drops the ball. If the safety experts input contradicts popular belief, production or efficiency, their views are often debunked, and they are scolded. When human nature finally catches up to the guru and they err in judgement, they are blamed, shamed, and often demoted or replaced. They are tarred and feathered when they come up short while trying to predict and prevent events. With divinity comes a great amount of baggage. The safety god can never admit they are unsure, that they are unknowledgeable of a certain area, or that they might have been wrong, for that would near instantly remove their infallible celestial status. So, what really happens? Safety experts shoot from the hip. They lean on what they know, try to learn what they do not, and then make up the rest. In a world of fast and easy answers, the organizations that safety professionals serve expect nothing less. The safety professional has become the "fast and easy button" for many organizations, a singular and highly visible point for answers, action, and blame.

Why do safety practitioners lean into the role of guru, god, and expert? It feels great! It feels good, at least for a while. We are looked to by executives and frontline employees alike, for our all-knowing wisdom and input. We grow more and more comfortable in the role of deity. The lashings, although painful and humiliating, all seem much more worthwhile when accompanied by the feeling of being needed. Our egos kick in and we begin to crave and promote our elevated status. Sometimes, when at its absolute worst, practitioners begin to honestly believe that they do indeed hold all of the answers, that only they know what's best, that the ideas of others are somehow beneath their own, and that they are in fact the "great knower of safety things!" It's a dangerous, yet easy, trap to fall into. We desire the feeling of being needed, we yearn to be highly regarded, and we pray that we are a necessity.

In an ever more cost saving and efficient world, safety practitioners often feel it necessary to continually justify their existence. Through dependance, we hope to create necessity for the practitioner. We find ourselves back at the question of value; we believe that being a "grand knower of safety things" makes us invaluable to our companies. If we hold all of the required answers, information about the most sacred of areas, then we surely cannot fall into the category of fat to be trimmed. If we are not gurus or we are discovered to be unknowledgeable of certain obscure areas of occupational safety and health, then we might suddenly seem much less valuable to those that employ us. If we cannot produce instant and easy answers on demand, about anything and everything safety, will we find ourselves on the fast track to the unemployment line? Everyone, safety practitioners included, craves certainty and stability. Fear is a painful, yet powerful motivator. We fear that if we do not live up to the divine standards of safety god, then we will swiftly be replaced by someone that will. At the very least, we will be replaced by someone that pretends that they can.

Safety practitioners and the companies that employ them, fall for the myth of the safety professional. They lean into the notion that the practitioner should be *all-knowing, a selfless sacrificer, a soothsayer and predictor of accidents, and fixer of company woe.* From this definition grows the guru. Our shared flawed beliefs about what a safety practitioner is, and what they should do, creates the guru-class of safety professionals. The environment that industries have created around the professional practice of occupational safety and health, is finely tuned to produce a steady stream of safety gods. As safety professionals, we regularly lean into this idea. While most do not classify themselves as Christlike, outright. Many often describe themselves as "safety experts," "SME's," "thought leaders," "masters," or give themselves other neat self-proclaimed titles that promote their all-knowing status. The profession has indeed played a key role in creating and perpetuating the legend of the safety god, but we did not do it alone. Companies, Industries, and people in general, all desire a guru to turn to for solutions to the things that plague them.

A quick appraisal of everyday life will reveal the emergence of gurus and gods in practically everything. Do you need to lose weight? There's a guru for that! Do you need some parenting advice? Guru! Do you want to start a podcast? You guessed it, guru! Its human nature to want easy, efficient, and miraculous cures to the things that cause suffering in our lives. The guru often derives their authority from the notion that only they hold the cure, or they set themselves apart as an example of what one could aspire to become by simply following their mystical teachings. According to a 2018 study released by the *Centers for Disease Control and Prevention*, nearly half of Americans are actively attempting to lose weight. Upon further investigation, you will find that along with this effort comes rampant use, and sometimes abuse, of prescription and over-the-counter diet pills. It doesn't take much looking in this area to discover a long list of dietary experts, chiseled and oiled up gurus that are more than happy to sell you their very own brand of weight-loss supplement.

Safety gurus are not that different from their fitness and diet counterparts. They have created a business from so-called "miracle cures" and are the self-proclaimed sole source of easy pills and answers. The safety guru holds the answers, and only they possess the cure. The infallible Christlike safety god is what you should aspire to be like, you need only to listen and follow in their footsteps! Oh, and buy these diet pills for the low, low price of $49.99! Companies have a source of pain, hurting people at work, and they desire an easy, efficient, and miraculous cure to end that suffering. While organizations create the demand for the guru class of safety practitioner, safety practitioners are the ones that willingly accept the title and the challenge. Together, we continue to insist on the existence of, and dependence on, safety gurus, gods, and experts. Do you have safety problems? There's a guru for that!

Now, I think that it should be said that this not some battle cry against seeking understanding, expertise, or know-how. It's also not meant to belittle the powerful and deep understanding and knowledge bank that many practitioners possess. In fact, it's quite the opposite. Let's reflect back on some of the definitions that we recently explored. Claiming expertise, simply put, means that you have mastered a subject. You have achieved a level in which you now know all that there is to know about a particular subject or skill set. When we claim to be a "safety expert," we are saying to ourselves, and to the world, that we have learned all that we will ever need to learn. We now know all that there is to know about safety. How profoundly stupid is that? A safety guru, god, or expert needs not to seek out deeper understanding, expertise, or know-how, because they already know all that there is to know. The most knowledgeable and effective safety practitioners are the ones that can openly admit that they do not possess all of the answers, not the ones that claim that they do. These are the practitioners, the ones that despite their enormous well of know-how and knowledge, that can approach work with an open curious mind and learn from those that get things done.

There are deeper and more harmful results, yet. We create safety dependency, rather than autonomy and safety innovation. We force less engagement and involvement in safety by sole sourcing answers from the safety guru. We choose to not learn from those that get things done, opting to learn from those that have a book that tells them how people get things done. The safety guru dangerously hoards vital safety knowledge and information, dispensing it back to the organization on demand and upon request. We consolidate our safety knowledge bank into a few "great knowers of safety things," and then demand that our employees always "go ask safety" before proceeding, in an effort to prevent bad things from happening. Rather than leaning into diverse thought and opinion from across our organizations, we have created a single point of failure, the safety guru.

In keeping with the understanding that people want easy, efficient, and miraculous cures for the things that suck, we will be brief. There is no room in our world for safety gurus, gods, or experts. Together, safety professionals, companies, and industries must lean into the idea that we get smarter by bringing together diverse groups of people in open dialogue, not by singling out and depending upon a small group of "great knowers of things." Organizations should encourage safety innovation and creativity from those that accomplish work and should learn frequently and deliberately from them. They should view the role of the safety practitioner, not as guru or god, but as a facilitator, conversation starter, team member, and team builder. As safety professionals, we must let go of the ideas that we possess all of the answers, that our expertise somehow outweighs the practical knowledge of those that accomplish work, and that by being a guru we are adding value. We must embrace our newfound role as facilitator, innovator, conversation starter, team member, and team builder, and we must always use our deep well of knowledge with the understanding that it will never be deep enough.

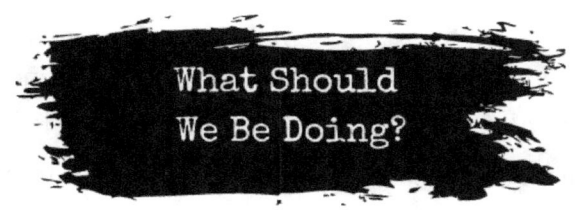

What Should We Be Doing?

What should we be doing? A simple question that safety practitioners, young and old alike, have asked since the very beginning of the profession. Due to the *"junk drawer effect"* that has already been mentioned, the practitioner often finds themselves doing what seems like everything. Safety professionals are frequently left holding a mixed bag of safety this, and safety that, along with anything else that could even remotely be classified as "safety" anything. Practitioners are tasked with everything from industrial hygiene to home safety initiatives, with compliance to managing and maintaining a small warehouse of personal protective equipment, with fatality prevention to ensuring that the restrooms have an appropriate amount and ply of toilet tissue, and on, and on. The safety profession has indeed succumbed to *the junk drawer effect,* becoming the home for anything and everything that organizations do not know where to put elsewhere.

This chapter started with a simple question – *What should we be doing?* But let's start from a different position, let's ask a better question in order to work towards a better answer: *What really matters?* As any professional in any field, the amount of time we have available to invest in any given item is finite. We must invest our "give a damn" wisely, we should invest our time into things that truly matter. While this book cannot give you a pretty and neat list of everything that should really matter to those that professionally practice safety, here is a starting point for how to think about that question:

What is the Shit that Kills People?

What Can Make Work Suck Less?

What Adds Value?

What Promotes Learning?

What Creates an Environment in which Honesty is Possible?

In the world of professional safety in particular, we often find ourselves aggressively focusing on the wrong things. A thorough examination of *what really matters?* is vital to ensuring that we are making the wisest choices possible with our efforts. It's extremely easy for the safety practitioner to fall into the trap that many organizations have set for them, this idea that says: *"If it's even remotely related to safety, it's super important and must be treated as such."* That trap, while seemingly so simple and morally sound, leads our profession to treat toilet tissue quality and availability with the same reverence and focus as fatality prevention. The trap of *"everything in safety matters, and it matters a lot!"* leads us to completely abandon prioritization and miss the mark on *The Shit That Kills You (STKY)* and *The Shit That Really Matters (STRM)*.

Prioritization matters, and it matters a lot! Unfortunately, it seems as if we have completely avoided the subject. We have continued to rally behind this idea that everything in safety is equally as important, which has resulted in us only wasting time and creating more safety bureaucracy and headache. A healthy dose of reality and prioritization is needed to get us out of this hole. A great place to begin to find that reality, and a good starting point for building a more effective safety "hit list," is to strip away the things that simply do not matter. Now, with the random corporate "go-do's," the awareness campaigns, and other fluff and garbage stripped from the list, we can begin to give our safety "hit list" some order by sorting by level of importance.

It should be said that some important items that fall into the category of Industrial, Health, and Safety, simply matter much less than others. These are the often required and necessary things that are probably now flooding your mind. They are the things that will always consume some portion of our time, and rightfully so. As an example, compliance in general, while important, should never be placed above, or in line with fatality prevention. That aggressive companywide driver awareness campaign, the one that you just invested hundreds of manhours into to tell people to "pay more attention," should never be placed above finding and fixing flawed or missing barriers if it should even exist at all. But that is the position that we regularly find ourselves in as safety professionals; we spend our days aggressively focusing on the wrong things. For good measure, we will ask the question once more, *what really matters?* What matters, a lot? What really, really matters? What are the most valuable and impactful areas that we should be investing our time and company resources into? If we answer that question honestly, the fluff and the bogus safety campaigns naturally begin to fall away. The important, but less important things, begin to drift towards the middle, and the *Shit that Really Matters* floats to the top of our safety "hit list." If it doesn't really matter, why are we focusing on it anyways?

We've started with, and continued to use, a safety "to-do" list that has had only one bullet. A bullet point that stated, "everything matters, and it matters a lot!" But with a strong dose of prioritization and a laser like focus on what's truly important, we can begin to expand that singular item into a viable and prioritized list of the things that really matter. We can strip away the clutter from the safety junk drawer and can spend our time aggressively focusing on the right things. Through prioritizing the *Shit That Really Matters*, we can begin to make sense of the chaos and rally around the things that really protect workers lives. Without the useless junk, we can focus on facilitating learning, we can invest our time into creating things that add value, we can focus on connecting people, we can promote an environment that yields honesty, and we can spend our days making work suck less. But as long as we continue to fall for trap, *"everything in safety matters, and it matters a lot!"* we will be doomed to a life of tail chasing and safety mediocrity.

Now what? Developing a safety "hit list," one that revolves around *The Shit That Really Matters*, is only half of the battle. Now that we have applied levels of importance to the things that we focus on, how should we approach those things? We have now identified some of the most critical areas that impact our workforces and must recognize that a poor or ineffective approach to these things can end disastrously bad. How should we approach these vitally important items, safety in general, and practically everything else that we touch or seek to influence? If we want to influence them positively, there are two simple rules that should always apply:

Start from a place of trust, rather than from a place of distrust

Do things with people, rather than to people

Trust should be our neutral position for both ourselves and the organizations in which we work. We regularly start from a place of distrust of employees. Procedures, rules, guidelines, check sheets, and other familiar artifacts of dated safety management techniques are often written from a starting point of not trusting employees to make the "correct" decisions or to do the right things. Then, when something inevitably goes wrong, human error happens, or we have an operational surprise, we double-down on distrust with more rules, harsher punishments, and heavier oversight. Nothing positive can grow from this misanthropic starting position, we only further degrade trust and belittle our employee's contributions to organizational success. To change our starting position from distrust to trust, we must shift our underlying assumptions about people in general. We must begin to lean into the goodness of people, we need to understand that people do things that make sense to them in the moment, we must believe that those at the coalface hold the answers that we so often seek, we must know that the vast majority of people we encounter act only with the best of intentions, and we must truly accept that human error is so normal that it is boring. Trust is a commonly sought state for many organizations, but few seem to achieve it. Often, companies seek to create this desired state by simply asking for it: *"Pretty please, trust us!?"* But actions always speak louder than words. If our desire is for employees to trust up through the organization, we must first be willing to genuinely trust down. Only then will we create environments in which honesty is possible. When trust of our employees is our normal "organizational neutral," we can begin to do safety with our employees, rather than doing safety to our employees.

Keeping in line with our typical misanthropic views of those that we employee, we frequently feel the need to do things to them, rather than including them and viewing them as valuable contributors. Our persistent belief that those nearer to the sharp end of the work are somehow less than those nearer to the blunt end, that they must care less than us, or that their input is somehow less valuable, leads us to believe that workers are dumb and must be told what to do at every turn. Nothing could be farther from the truth. People are the solution, and they are the subject matter experts of their jobs. People make successful, safe, and efficient work happen, each and every day, while dealing with a multitude of external and internal pressures and conflicts. If our desire is to deeply understand how work actually happens, if we hope to create more robust and resilient systems, if we desire lasting and positive change within the complex sociotechnical systems that people work, if we truly value learning above all else, then we must embrace workers as the solution. We cannot simply do things to people, even if we vehemently believe that we are acting in their best interests, if we hope to create valuable, meaningful, and lasting positive change within the companies that we serve. People create safe outcomes, on-the-fly, and in real time. Our role should be supporting them in that, not in pretending that we create safety for them in some shiny conference room in an office far, far away.

Now, I'm not saying that all of our efforts are futile attempts that will always fall flat. But they will be, if we do not involve those that operate day-in and day-out within the complex sociotechnical systems that we seek to influence. If we hope to create true betterment, we must first admit to ourselves that we do not always know best. Once we can stomach that sometimes hard to swallow fact, we must listen to, learn from, and involve those that actually get things done in our workplaces. Our contributions do add value, an immense amount of value, when we start from the right place. If we embrace a better definition of our role, a definition that views the safety practitioner as a communicator, facilitator, innovator, team member, and team builder. As a curious person with an obsession for learning about work, and an evangelist for organizational betterment, we naturally begin to drift towards starting from trust and doing safety with our employees. But, if we continue to live the role of an all-knowing guru, a know-it-all, a soothsayer and predictor of accidents, and fixer of company woe, no matter how well intentioned our efforts are, they will only lead to struggle and increased safety work.

We now find ourselves at point of shifting our efforts towards the really important things, a move towards investing more and more of our time into *The Shit That Really Matters*. But we still have a truck load of random tasks and "go-do's" that we must accomplish on a daily basis. We still have compliance this and corporate initiative that, we still have classes to teach, and various bits and bobs of company safety management systems to oversee. How do we make it all work while still keeping the really important stuff at the top of the list? Personally, I like to think of all of those items as sliders on an audio console. If you push them all to 100%, you will quickly be deafened by squealing feedback. If you set them too low or turn them off completely, nothing happens. It's all about finding the best mix for the particular situation you find yourself in. Maybe we set our focus on compliance to a solid 4, we turn off awareness campaigns completely, we tune our focus on lagging indicators to 1, and we crank our focus on *The Shit That Kills You (STKY), The Shit That Really Matters (STRM)*, and Trust to 11! While this metaphorical picture demonstrates something that we have already discussed, it paints a clearer portrait of how that actually happens. Our focus can and will shift, sometimes this will happen from day to day or hour to hour. But those things that actually matter, and matter a lot, should be cranked to the max and the knobs should be broken off there. As with our audio mixing console, we have a limited amount of real estate and capacity. If something that adds little to no value is taking up a channel, it should quickly be removed and replaced with something more worthwhile. If we turn down *The Shit That Really Matters* so that we have more capacity for things that do not, the outcome of our mix will be rather unpleasant. If we try to crank everything to 11, because "*everything in safety is equally important,"* our efforts will be rendered effectively useless, and we will find ourselves burned out and frustrated with the resulting ear-piercing feedback. A world-class audio mix is often defined as having everything in its rightful place and set at optimal levels. That sounds like a valuable concept that could easily apply to how we prioritize and tune things, both in our daily lives as safety practitioners and within our safety management systems.

With our newfound focus on *The Shit That Really Matters,* an understanding of how to excavate *The Shit That Really Matters* from our cluttered safety junk drawer, a heavy dose of prioritization, and a better lens in which to approach these vitally important things, our safety "shit list" is evolving towards a workable safety "hit list." Through our shift in thinking on how we prioritize and approach our safety focus areas, we are left with a clean, simple, and actionable list of *The Shit That Really Matters*. A streamline list, one that we can now approach from a mindset of betterment, rather than from a place of blame. A question, one that seems so simple on the surface, still remains unanswered: *What should we be doing?* As a safety practitioner, what should our day-to-day job look like? With the particulars of what really matters and overall approach out of the way, let's talk about what the daily role of the safety practitioner should look like.

To get to the answer of *"what should we be doing?"* Let's first start with where we are currently at. Those that find themselves in the professional practice of safety will typically find their days filled to the brim with the useless junk that we have already touched on. This junk, trash that is regularly sold as treasure by our companies, peddlers of "easy button" safety programs, and sometimes safety professionals themselves, engulfs our daily existence. We often spend our days conducting *Behavior-Based Safety Observations*, generating observation trending, compiling data about all of the bad things that have happened, doing endless compliance audits, using these extremely flawed data sources to attempt to predict bad things on the horizon, and telling people to "be more safe" or to "pay more attention" to prevent said bad things from occurring. In addition, we find ourselves managing anything and everything that has crept its way into our space; we live inside of the safety junk drawer. When we are done auditing the stock of toilet tissue in the restrooms, doing our sacred observations, and have completed our vitally important awareness campaign on *"The Dire Hazards of Coffee Making in the Workplace: Will You Be the Next Coffeemaker Fatality,"* what's left? Quitting time, a wheelbarrow full of frustration, and hopefully an ice-cold pint at the bar. We spend our days aggressively focusing on the wrong things. In part, because of the way we have traditionally defined the role of the practitioner. As previously discussed, we regularly define the role of the safety practitioner as *"an all-knowing guru, a selfless sacrificer, a soothsayer and predictor of accidents, and fixer of company woe."* With that definition, is it any surprise that we find ourselves rendered useless and are left burned out and frustrated by our efforts? We spend our days aggressively focusing on the wrong things, and we know it.

So, *what should we be doing?* Let's reflect back on the better definition of the role of the safety practitioner: *A communicator, facilitator, team member, and team builder. A curious person with an obsession for learning about work, and an evangelist for organizational betterment.* Each little piece of that definition falls into the category of *The Shit That Really Matters*. Our days should be consumed with a focus on learning and growth, a focus on betterment, a focus on making work suck less, a focus on *Making the World a Better Place to Work*, and ultimately, an extreme and unwavering focus on *The Shit That Kills You (STKY)* and *The Shit That Really Matters (STRM)*. What should we be doing? We should be spending every ounce of our precious time trying to put an end to workplace fatalities and life altering accidents. We should spend our days determined to make things better, and to make work suck less. We should be doing this, not by doubling down on blame and compliance, but by facilitating learning, encouraging creativity and innovation, by learning from normal work, encouraging micro-experimentation and worker adaptability, and growing environments in which honesty is possible. We should invest our careers into working on *The Shit That Really Matters*; we must aggressively focus on the right things. Spending very much time on anything else, is just a waste and a move away from betterment and growth.

As you now find yourself reflecting on how you spend your days as a safety person, and pondering on the question, *"what should we be doing?"* I hope that a few key items come to mind. What really matters?

What matters the most? Out of all of the clutter in our safety junk drawer, what is trash and what is treasure? I hope that you find yourself sorting through and prioritizing by *The Shit That Kills You (STKY)* and *The Shit That Really Matters (STRM),* all while cleaning out the clutter and taking out the garbage. With our better definition of the practitioner, a meaningful and prioritized safety "hit list," and a better approach, we should be able to begin to make sense of the clutter and chaos that has become the professional practice of environmental, health, and safety. So, when we ask *what should we be doing?* The answer is simple, yet powerful. We should be spending our days aggressively focusing on the right things, we should be investing our time and "give a damn" into the things that actually matter.

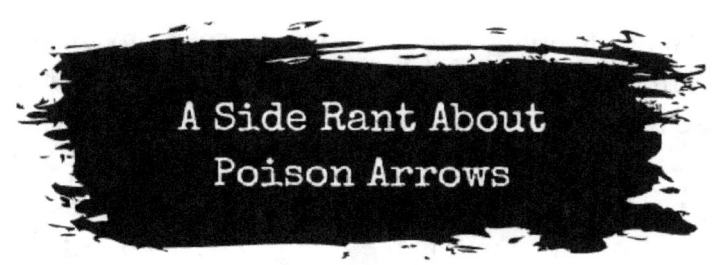

A Side Rant About Poison Arrows

We have spent a good amount of time talking about the *Shit that Really Matters*, what we should be focusing on, and much more. But, due to many of the underlying assumptions that swirl around safety at work, we never quite see those ideas become reality. Our organizations toil away tirelessly and aggressively focusing on the wrong things, often creating more harm than good with their well-intentioned efforts. Our assumptions about worker safety, how we impact or influence it, and the role of the safety practitioner, has led us astray. We have been led down a path of focusing on the trivial, of asking unimportant questions, and of seeking organizational enlightenment through the pursuit of knowing more about the insignificant. We waste our efforts, our time, our questions, exploring the meaningless. We collect and analyze these pieces of *fool's gold*, hoping that they will one day reveal to us the secrets of pre-accident prediction and prevention, but they never do. As Buddha once said, *"Better than a thousand useless words is one useful word..."* To start focusing on doing *Safety Better*, to seek any form of betterment for that matter, start by asking better and more meaningful questions.

Contained within the *Pāli Canon*, a collection of scriptures in the Theravada Buddhist tradition, is the story of the poisoned arrow. Buddha, while replying to a young impatient student's inquiries about the afterlife, shared with him a story about a man that was shot with a poisoned arrow. As the mortally wounded man laid upon his deathbed, he refused his family's offers to bring medical help. The man did not desire medical intervention, he desired answers. Before he would allow his family to summon the doctor, he wanted to know who had shot the arrow? What type of person was he? What was his height and strength? What was his skin tone? What was the composition of the bows sting? What type of bow was it, and what materials was it constructed from? As the man wondered about the arrow's feathers, whether they came from a vulture or peacock, the man succumbed to his injury before getting an answer to any of his questions.

As absurd as the man's course of action appears, it sounds eerily familiar, huh? We often follow a similar path within our organizations. We waste valuable time, resources, and focus on pursuing answers to questions that do not really matter. We aggressively focus on the wrong things, and it shows. Post event, we pour all of our collective energy into asking things like "what did the employee have for breakfast?!" or "is this a willful violation, reckless behavior, or an unintended error?" or "what is the involved employee's astrological sign? Is Saturn in retrograde?!"

We work diligently to compile our trash-heap of useless data, cherry-picking and force fitting the useless bits into a neat and clean incident report. After our dutiful examination of the "facts," we now have our culprits. Through our efforts, we have discovered the cause of our pain! The employee did not have a proper breakfast before coming to work, and Saturn was indeed in retrograde. The company has a poster that talks about the importance of consuming a nutritionally balanced breakfast before coming to work; the employee simply chose to not eat! Surely the employee knew Saturn's status, and should have known how it could potentially impact their particular astrological sign! Willful violation it is!

After our atrocious investigatory efforts and an expedition into blame, we then excavate through our annals of horror, our incident log, to seek out evidence of a "trend." We bend, mold, fit, and interpret this historic information in a fashion that conforms with our hypothesis. "Ah ha!" we think to ourselves, already patting ourselves on the back for our handy detective work. "We have four events that resulted in *OSHA* recordables over the past 100 years, all of which the involved employees had the same astrological sign and had not eaten breakfast! A clear and demonstrable trend!" we exclaim as we rush to get these epiphanies into *PowerPoint* before they fade from our minds. Safety sherlock strikes again!

From our worthless adventure as safety detectives, comes even more worthless "fixes." We will now spend months of our time constructing, rolling out, and enforcing our brand new (and corporate approved) safety campaign, *"Eat breakfast like your life depends on it, because it does!"* We will force feed our workforce lengthy presentations chocked full of the perils of not eating a balanced breakfast before arriving at work, pointing to our newly discovered "trend" as evidence of its importance. We will spend thousands of dollars on *"Eat Breakfast or Die!"* banners, posters, hardhat stickers, and t-shirts, to make sure everyone understands how important the subject is to the organization. We will add *"have you eaten breakfast?"* to our pre-job briefings and will implement *"breakfast checks"* at the entrances of our facilities to double-check that people have eaten before arriving on site. We'll parade around the unfortunate involved soul, the one who's "willful choice" to not eat breakfast "caused" the event, on a roadshow to tell their tale about the dangers of not eating breakfast before work and to "teach them a lesson!" We will include their story in our safety meetings and at the beginning of nearly all safety-related presentations, memorializing the event forevermore. Management and safety will do spot checks and focused observations, increasing the organizations oversight of this crucial area of concern. From these observations we will compile and trend data to measure our compliance with our new *"eat breakfast or else*" rule, and to examine the performance of the program. When non-compliance or a decline in performance is discovered, we will quickly course correct by retraining our employees, beating them harder, or offering them incentives for increased performance. Worse yet, we will mistake this misguided boondoggle for learning and bask in the feel-good sunshine of our efforts.

Our aggressive focus on the wrong things, our pursuit of unimportant information, is not limited or contained to our post-event worlds, it permeates our day-to-day operations. More often than not, it does not require some grand operational surprise or upset to send us down the path of seeking out answers to unimportant questions. The over focus on prevention, one that grows from our belief that says, *"we prevent big bad things from happening by preventing little bad things from happening,"* leads us to do some pretty wonky things in the name of zero. Just as the man in Buddha's story, rather than focusing on what really matters, we waste our valuable time and efforts on things that simply do not matter very much at all. Rather than asking meaningful questions and seeking out valuable learning, we pursue useless KPI's, data, measures, and metrics.

We parade this data and faux learning around as the gospel, sacred information that with enough effort will finally allow us to prevent all bad things from happening. As organizations and as safety practitioners, we have become enamored by so-called "leading indicators." These bits of organizational data that promise to grant us the superpower of predictive capacity, quite literally allowing us to peer into the future and prevent events from occurring. We lay on our deathbed after being shot with a poison arrow, seeking answers about compliance, the number of pre-job briefs people have completed, how many times people used *Stop Work Authority* last month, and observation data trends, rather than asking for medical intervention. We expend our finite resources and time on observing and correcting unfavorable behaviors, labeling people's behavior and actions, crafting nifty safety campaigns from the resulting "data," asking dumb questions like *"why do our employees not care enough?!"* responding to less than favorable data by saying, *"If people just followed the rules..."* and we often do this up to our employees dying breath. We have become so obsessed with knowing everything, that we know nothing. At the very least, we do not know enough about the important things, and it's killing us. We do not know enough about the *Shit That Really Matters*, because we are using all of our time and resources focusing on the shit that does not. Even worse, we believe wholeheartedly that through our pursuit of the trivial, not through the pursuit of the meaningful, we will finally find the answers and the fixes that we so desperately seek.

An even scarier thought yet, due to our focus on questions and answers that do not matter, our poor reactions to operational surprises, and the horrendous "corrective actions" that grow from these, maybe we have driven people to simply stop reporting the "little things" all together? So, the "little things" might have not actually been reduced at all. Our obsession with knowing and analyzing all, could very well be leading to us knowing much, much less.

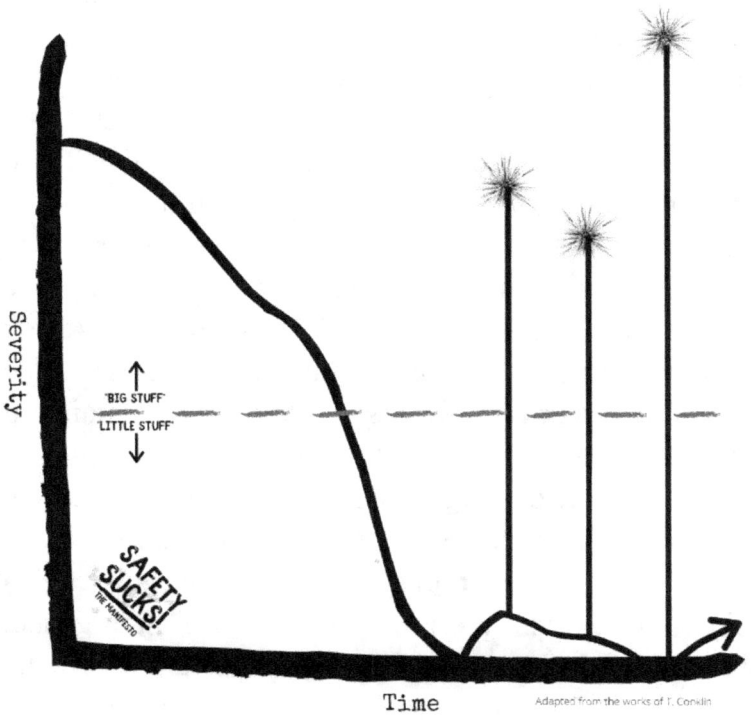

A hard pill for our organizations to swallow is this, not everything matters and not all information is useful. We have become obsessed with this notion of measuring our way to success; we have become addicted to data. We often value it, and its collection, above the care and wellbeing of our people. Nearly everything that we do within our work worlds must contain some numerical indicator to measure success. Our bloated and swollen event reporting systems have grown to page after page of useless trivial questions, data points for the organization to collect and trend. Rather than asking employees meaningful question and soliciting their feedback, we demand rigid black and white measures for anything and everything. To not measure, measure softer, or to measure differently, is corporate blasphemy! We can no longer waste our time on things that are unimportant. We find ourselves back to a key and repetitive point of this book, a focus on the *Shit That Really Matters.* We must maintain unwavering attention on that which is truly worthwhile, for the good of our organizations and our people. We must break the cycle, our repetitive focus on relying on the trivial, the unimportant, and the meaningless to guide our efforts. We must seek out the intelligence, and the wisdom to separate the important from the unimportant. We must ask better questions if we seek more meaningful answers. Knowing the difference between what is treasure and what is fool's gold, can make the difference between overcoming a difficulty or being overcome by it.

The Safety Dose-Response-Harm Relationship

As companies, industries, and as a profession, we have long ignored the challenges of safety in practice. We have been seduced by "easy button" approaches to safety; we have attempted to cure complexity through simplification. We have tried and tried to turn nonlinear chaos into linear order through a plethora of "simple" fixes. These easy and "actionable" approaches have failed us time after time. Yet, every time that they do, we double-down on their use. Even worse, through the application of "easy button" safety, we have crafted a work world full of bureaucracy and developed an extreme focus on things that don't really matter. Our desire to simplify our complex worlds into clean and linear order, has only served to complicate them.

Complication, as defined by *Merriam-Webster*, is something *consisting of parts intricately combined*.

Our safety management systems, the approaches that we take to worker safety, can easily be defined as safety complication. We take that which is the most valuable of all, the lives and wellbeing of people, and we complicate it beyond belief. We layer and interconnect procedures, we confine our workforces with rigid prescriptions on how work should be done, and we make it harder to "be safe." We write rule after wonky rule, demands that often say something like, "thou shalt always wear high-visibility clothing unless its Tuesday, then just a vest is ok. But on Sunday, you need a bright-green hardhat, too. Because, you know, safety!" Within our endless rules, procedures, and checklists, we leave a trail of breadcrumbs that lead to other rules, procedures, and checklists. We create complicated and unique processes and programs, ones that must be used together and executed flawlessly, to prevent bad things from happening. Often, in an attempt to address the stuff that kills us or the stuff that really matters, we draft useless documents that are hundreds of pages in length. We hide and drown out vital bits of information about well-known killers in a sea of compliance this and "cover your ass" that. Rather than approaching complex and dangerous work by seeking clarity, we create more and more complication. Complication, the creation of more meaningless work and bureaucracy, that figuratively and literally cripples our employees.

We arrive at these conclusions about safety due to our firmly held underlying assumptions about what safety is, how we define it, and how we do it better. These deeply rooted beliefs, assumptions that we have already discussed, lead us down a path of easy complication. As mangers, leaders, and safety practitioners, it really feels like we're doing something worthwhile, it feels as if we are making an impact. We are indeed impacting our workforces, just not in the positive ways that we had hoped for. The list of unintended consequences, this catalog of good-hearted efforts, is long and frightening. As disastrous and harmful as these unintended results have been, we seem to ignore them and the ways that they impact our organizations.

Our efforts have been well intentioned, we have pursued them for the best of reasons. Our hearts are in the right place and we desperately seek to better safety. But as long as we are starting with flawed assumptions and beliefs, we will never realize our goal of *safety better*. In fact, we will regularly find ourselves doing safety worse. I want to draw your attention back to the *Dose-Response Relationship* that we touched on earlier. We regularly "overdose" on safety, but we never realize it. We tell ourselves, "if 100mg is good, then 1000mg will be great!" We draw an assumption about safety dose, one that says, "*the bigger the dose, the better the result.*" In our minds, it looks like this:

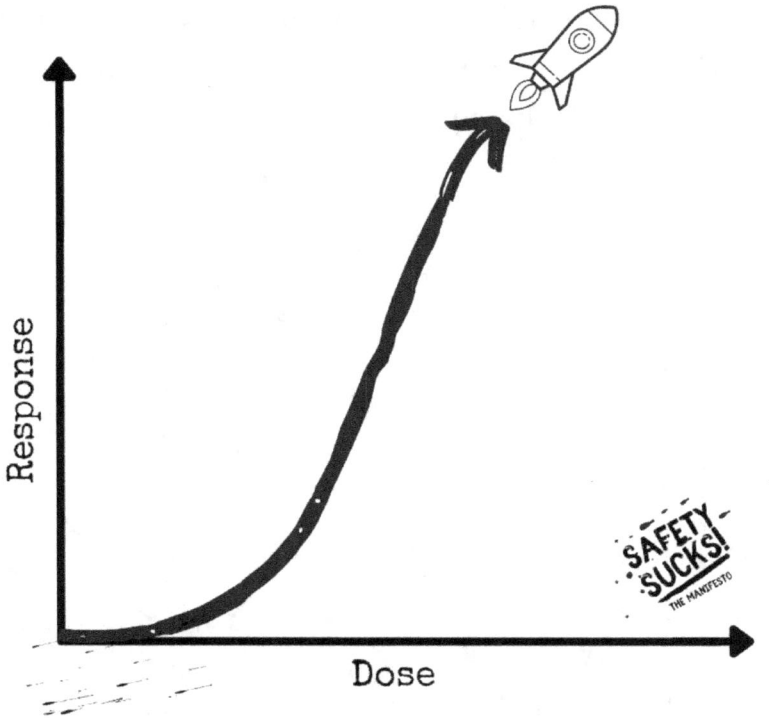

We hold fast in our belief that by continually increasing the dose, eating more and more easy to swallow pills chocked full of "the ways we've always done things," that we will eventually blast-off into world-class safety performance! But as we are so painfully aware, we never seem to have that "blast off" moment. Simply put, this is because more dose does not always equate to increased response. In fact, especially as it relates to safety dose, more dose frequently equals wasted time and less response. With safety dose, the *Dose-Response Relationship* looks more like this:

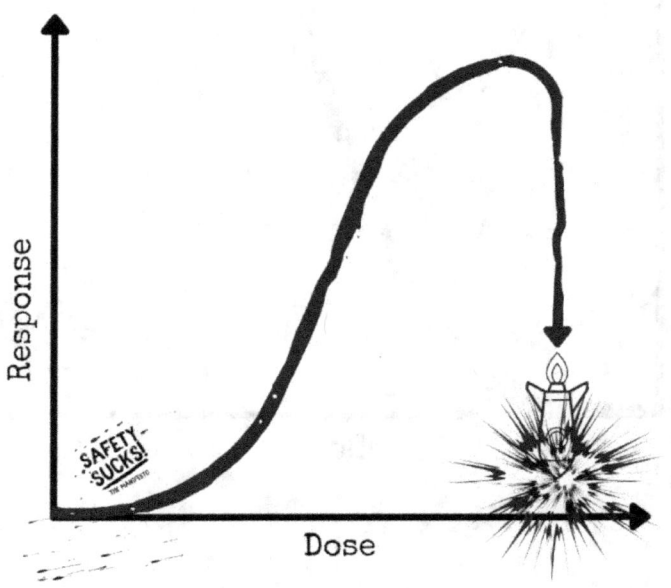

In all actuality, this constant increase of dose is regularly accompanied by a significant dose of harm. As someone so eloquently once said, *"the difference between what is poison and what is a cure, is dose."* Our non-stop dose escalation eventually creates more poison than cure. At some point we overdose, and a long list of negative side effects follows suit. We dose and dose well beyond the point of harm, all the while we pat ourselves on the back for our efforts. Even as we begin suffering from the negative side effects, we just can't seem to stop. We have become "safety addicts." We hold true to our belief that, *"everything in safety matters, and it matters a lot!"* We couple this belief with another we have already discussed, *"to get better, we need to do what we have always done, just harder!"* From these, we conclude that by doing more of the same harder, and continually doing more of it, we will eventually reach "blast off!" Rather than blast off, our progress stalls and we ultimately create harm. We funnel "easy pills" into our safety management systems, and never come to the realization that we are making things worse. Here is one more example to paint an even clearer picture of the *safety dose-response-harm relationship*:

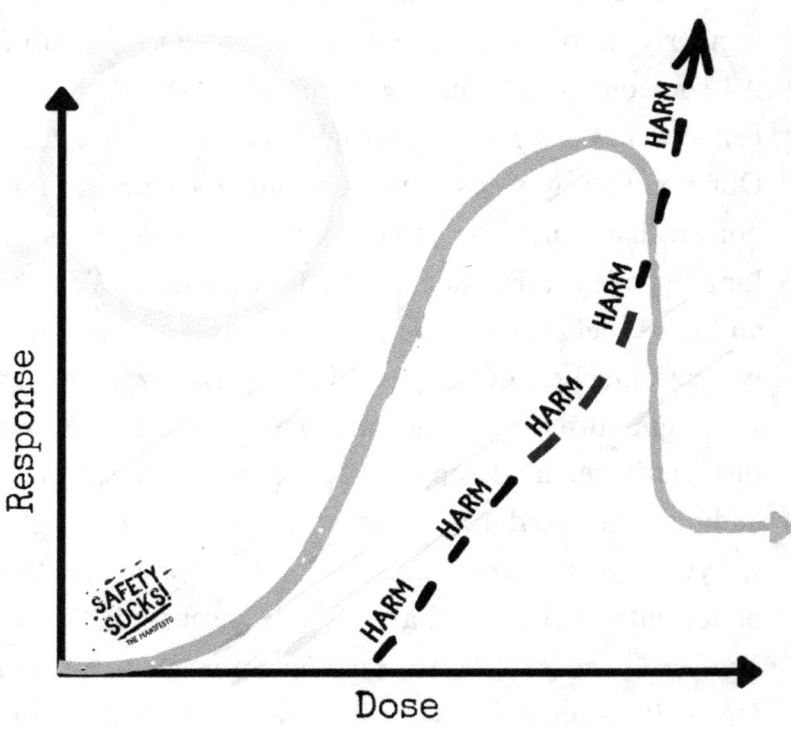

Now, with an understanding that our commonly held beliefs about safety are not the greatest, we should be able to admit that we are not starting from the right place to begin with. Rather than a focus on the *Shit That Really Matters*, we find ourselves aggressively focusing on the wrong things. We are administering an ever-increasing dose, but we are treating the wrong disease. We are pouring more and more morphine down the throat of the organization, but we should be administering *Benadryl*. Not only are we administering the wrong medicine and prescribing it for the wrong illness, but we are also dosing it well beyond its effective and safe limits. How can we help to make sure that our "cures" are actually curative? First, we must acknowledge a few uncomfortable truths:

SOME UNCOMFORTABLE TRUTHS...

Doing the same things, just harder and faster, fixes nothing

A bigger dose of "safety" doesn't always make you "more safe"

Overdosing on safety to increase safety performance has unintended side effects. In fact, it's poisonous

Can we finally put an end to this insanity? Can we finally admit that we're "safety addicts?" Can we finally let go of this notion that we improve worker safety by an ever-increasing dose of "the way we have always done things?" Can we begin to understand that this "overdose" has left us with a safety hangover? We have a raging hangover consisting of endless clutter, bureaucracy, and mediocrity. If our desire is to actually get better, we should. We cannot afford to continue to cling to these flawed beliefs, they have left us treating the wrong problems. We can no longer continue to up the dose of our "tried and true" medicine expecting a different result, it's poising our organizations and making nothing better. We must curb our enthusiasm for quick fixes, and we must take time to learn before we treat.

To begin, we have to shift our starting position, we must start from a place of better assumptions. These better assumptions, ones that we have touched on throughout this book, are vital in shifting our approaches to safety. These new underlying beliefs that say, people are the solution, learning is everything, safety is the presence of positives, error is normal, failure will happen, we better system output by working on the input side, and on, and on, must become deeply engrained within us as safety professionals and within the organizations that we seek to positively influence. From this better starting position, we can begin to construct a better approach to address the *safety dose-response-harm relationship*:

WHAT WE SHOULD BE DOING...

Focusing on the right things, things discovered through a healthy dose of learning

Growing an understanding that everything, even the stuff that feels great, has unintended side effects

Understanding that the best way to "fix" the right things and avoid unintended side effects, is by learning from those that reside in the systems we seek to improve

A "miracle cure" can quickly become poisonous if overdosed or used to treat the wrong illness. In the

world of occupational safety and health, we often find ourselves doing exactly that, poisoning our organizations. We are "safety addicts," dumping more and more easy pills down the company's gullet, all in the name of "improving safety." But it doesn't have to be that way; safety shouldn't be that way! The deeper lesson behind the *safety dose-response-harm relationship,* is a cautionary tale of unintended consequences. It is a warning against fixing without learning, against *doing the same things over and over and expecting different results*, and a stark reminder that too much of anything is bad for us. Everything that we do, every action that we take in effort to influence our systems, has unintended outcomes and consequences. The exploration of these unknowns should be high on our list of important things to focus on, and it should remain fresh in our minds as we approach "fixing" anything.

A Newfound Focus on Learning

We have now arrived at a major shift in how we think about safety, and the role of the professional safety practitioner. The views of safety as an enforcer of rules, all knowing guru, complicator of work, creator of bureaucracy, permission grantor, fortune teller, and preventer of company woe, will fade quickly with our shift in underlying assumptions around the practitioner, and worker safety in general. Our new and better assumptions, ones that lead us to believe that safety is not a game of outcomes and that the practitioner is not just a manager of outcomes, now drives us towards better organizational and professional values. We find ourselves with a new and hyper focus on learning.

Let's circle back to a couple of our new assumptions about safety in general. We no longer define safety as an outcome, safety is not some number or metric to be managed at best, and worse, manipulated. Previously, we assumed that the presence of accidents indicated a lack of safety. But we now understand that a) *Not all accidents are preventable,* b) *ridding our organizations of lower-level events does not prevent larger, more catastrophic events in the future,* c) *despite our best efforts, prevention will always fail at some point,* and d) *safety is not the lack of accidents, it is the presence of positives.* From these better assumptions we can derive:

FAILURE WILL OCCUR
Failure is always an option. By assuming for failure, we find ourselves seeking margin over prevention.

SAFETY IS THE PRESENCE OF POSITIVES
Safety is the presence of things that help create better outcomes.

LEARNING IS EVERYTHING
Learning is vital as it is the only real tool that we have in creating safety

With this shift in views about what it means to "be safe," also comes a shift in thinking around how the safety practitioner supports this mission. We will now find ourselves called upon to be curious explorers of normal work, evangelist for organizational and system betterment, innovators, and facilitators of learning, just as we should be. But how do we accomplish that? First, we must remember that 1) *workers are not a problem to control, they are the problem solvers*, and 2) *by deliberately learning from those that do the work, we are creating betterment.* There are near-endless ways to gain operational intelligence, methods by which we obtain knowledge about how people actually accomplish work. No matter what method you choose to obtain this critical information, all should be applied with these ideas in mind. We must always:

> *Start from a place of trust, viewing workers as the problem solvers*
>
> *Focus on creating environments in which honesty is possible*
>
> *Understand that those that actually do the work, have the greatest understanding of the work*
>
> *Stay curious about how normal work occurs*

With some basics out of the way, let's explore a few of the many operational learning tools that we have at our disposal as safety practitioners. How can we seek to understand normal work? How can we obtain this vitally needed operational intelligence? Here are few great options:

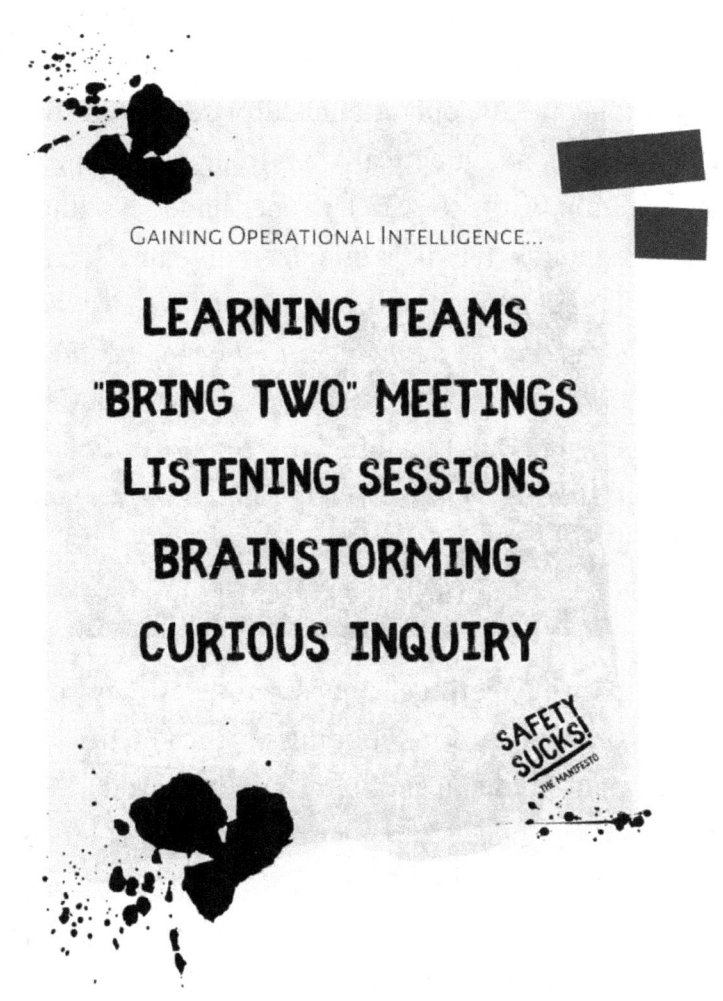

Learning Teams are one of the most powerful and easy to use tools that we have available us. WTF is a

Learning Team? A *learning team* is a facilitated group discussion that seeks to learn from everyday successful work, accidents, operational surprises, interesting successes, and practically anything else that the organization wants to gain a deeper understanding and greater perspective of. For a deep dive into learning teams, I would encourage you to read *The Practice of Learning Teams: Learning and improving safety, quality and operational excellence.* Sutton, McCarthy, Robinson. 2020. Along with *Bob's Guide to Operational Learning: How to Think Like a Human and Organizational Performance (HOP) Coach.* Edwards, Baker. 2020.

"Bring Two Meetings" are as simple as they sound and can be great for soliciting ideas to help solve shared pain-points or problems within a group. Prior to the meeting, each member of the group is given a problem statement and asked to "bring two" solutions or fixes for the given problem to the meeting. The ideas are then discussed and debated during the meeting.

A **listening session** is exactly that, listening to the workforce. It is similar to a focus group and it is a type of facilitated conversation with a group of people and is aimed at collecting information about their experience.

Brainstorming, whether used in a small or large group settings, is a method used to generate ideas to solve clearly defined problems or explore betterment opportunities. In controlled conditions and in environments in which honesty is possible, teams approach problems by asking "How Might We" questions.

So, what do we mean by **curious inquiry**? Simply asking goes a long way. It's simple, effective, engaging for all those involved, and something that we as safety professional are already tuned to do during the course of our daily work. The effective use of curious inquiry is one of those areas that can near-instantly change the dynamic that exists between the safety professional and the workforce. It is a great opportunity, one that exists at the point of "real work" in the organization, in which the practitioner can make a huge impact. Here are a few great prompts to use as you encounter work in the field:

Curious Inquiry Into Normal Work....

CAN YOU TEACH ME HOW TO DO THAT?
WHAT IS HARDER THAN IT SHOULD BE?
WHAT SUCKS ABOUT THIS JOB?
WHAT'S THE DUMBEST THING WE MAKE YOU DO TO WORK HERE?
WHERE DO THE RULES NOT WORK?
WHERE DO YOU HAVE TO MAKE DO?

SAFETY SUCKS!
THE MANIFESTO

No matter what tools that are utilized, we must have a firm understanding that learning is everything. Learning falls easily into the category of *Shit That Really Matters*, and we must maintain a hyper focus on deliberate and meaningful learning. As the legendary Todd Conklin stated, *"knowing less does not make you smarter."* Knowing more about how normal work happens, only serves to make us smarter, and smarter is better. Learning is the only real weapon we have in this fight, and it is the most valuable of assets to our organizations. As previously discussed, the shift in our definition of "safe" towards the presence of defenses, coupled with our understanding that learning is how we build, better, or fix defenses, leads to learning being valued by the organization. As safety practitioners our role now shifts to support this value, we evolve to become champions and facilitators of learning. Without deep, meaningful, and deliberate learning, learning that occurs in environments conducive to honesty, we will spend our days aggressively focusing on the wrong things and we will fail in our mission to *Make the World a Better Place to Work*.

The Care & Feeding of Safety Practitioners

The professional safety practitioner faces many day-to-day challenges that other professions do not. At the very least, practitioners are typically treated radically different than others within their organizations. As organizations, we hold firm in the idea that safety is super important, and rightfully so. But with this high value placed on safety, comes an extreme "importance" placed upon the role of the safety practitioner, one that is often misguided. Due to this "importance," we grow this idea that *"nothing can be left untouched by the hands of the safety practitioner!"* We then craft and shape an unrealistic set of expectations for the safety practitioner, we build a role that is primed for failure. But why? Organizations typically hold two strong, yet opposing assumptions about the role of professional safety:

-We assume-

Safety is everywhere, and so must be the safety practitioner*!*

-But-

Safety practitioners are costly and provide little value.

Our desire is to have it both ways; we *"want to have our cake and eat it too."* Our wish is to have a world in which the safety practitioner is deeply involved in every task and engaged with each and every employee, always! But we also strongly desire a world in which we cut safety staff to the bone, often only employing a singular or *"lone wolf"* safety professional, no matter the size and scope of the project or location. In one breath organizations preach about the "importance" of the safety practitioner, then in another they belittle their impact and cut their budget. They make ridiculous demands such as, *"95% time spent in the field,"* while also demanding that they never miss a meeting or fall behind on administrative tasks.

Allow me to paint you a picture, a real-life example, one that I experienced way back when. I was a *"lone wolf"* safety practitioner working at a large site that shall remain nameless. Allow me to define "large;" this site employed nearly 1200 employees during the slowest parts of the year. Now, relying on the assumptions about the safety practitioner that we just discussed, based on a typical 8-hour workday, what does that look like in real-life for a lone wolf practitioner?

24 SECONDS
Engaging with each employee daily not counting time to traverse the site, breaks, or other tasks

96 SECONDS
Engaging with each employee daily if we seek to only touch a quarter of the workforce each day.

216 SECONDS
Engaging with a total of 50 employees daily after allotting 2 hours for office tasks, 2 hours for meetings, and 1 hour for lunch and breaks.

9 SECONDS
Engaging with each employee daily after allotting 2 hours for office tasks, 2 hours for meetings, and 1 hour for lunch and breaks.

SAFETY SUCKS! THE MANIFESTO

With this example in mind, an example that I'm nearly certain you have also experienced or witnessed, is it really any wonder that safety practitioners find themselves worked into the ground? Is it all that surprising, based off of the assumptions that we just discussed, that companies and safety practitioners are left frustrated and wanting for more? Is it that shocking that organizations often demand that the practitioner work near-endless hours, many of which are unpaid, in an effort to better support the workforce? These harmful and deeply flaw assumptions long ago became basic job expectations for the professional safety practitioner. Expectations that safety professionals have embraced out of fear of unemployment and hopes of better serving our employees and employers. Expectations and demands that employers of safety practitioners now examine the practitioner against to measure their worth. The safety professional's duty to be the *selfless sacrificer* has become solidified, it is not only desired, but it is demanded.

From these contradicting beliefs stem a world of problems and frustrations, for both the practitioner and the organizations that they serve. For the safety professional, they find themselves being stretched beyond their limits, working themselves into the ground. Organizations feel perpetually let down by the practitioner for not meeting their impossible expectations of being everywhere and catching everything, all the while keeping up with any bit of administrative clutter that has found its way into the *safety junk drawer.* A dynamic is created between the practitioner and their employer, one in which the organization is constantly questioning the effectiveness of the safety practitioner. This *"what have you done for me lately?"* attitude leaves the safety practitioner on-edge and fearful. The employer of the practitioner is ever questioning and distrustful of the work ethic and ability of the individual that holds the position. This harmful dynamic is what leads safety practitioners to avoid their office, even when they have important tasks to complete. It causes them to feel the need to always appear busy, even in the slowest of times. It puts the practitioner in a position of constantly looking over their shoulder, with a well-prepared answer in hand, of exactly what they are working on during every millisecond of their day. This attitude drives the employer to constantly check up on the practitioner, double-checking and triple-checking that they are not slacking in their responsibilities. It causes the employer to micromanage the safety practitioner into the ground, and leaves the safety practitioner anxious, nervous, and feeling undervalued and belittled.

Now, I think it's safe to say that not all professionals are created equally, safety professionals included. This is in no way meant to paint all safety people as angelic beings that are devoid of the flaws of humanity. As much as we must lean into, and we must start with the underlying belief that the vast majority of people are good, we must also acknowledge that shit heads do exist. Most people show up to work every day for the right reasons, they make the best decisions that they can based off of the situations that they find themselves in, and they hope to do a good job. Shit heads are by far the minority in any profession that you examine. But even though they present as a minuscule amount, they do exist, and we must acknowledge their existence:

THE SHIT HEAD CLAUSE

There are bad actors that exist within our world. You always have the potential to encounter the ill-intentioned, con artists, crooks, the lazy, and sometimes the downright vile. Although rare, they do in fact exist. We must acknowledge their existence and be prepared for them. But we must always remember, the vast majority of people are not shit heads. Treat people like people, and deal with shit heads appropriately.

It's actually pretty simple, your safety practitioner is more than likely not a shit head. But if you do in fact employ a shit head, you should swiftly *"make them available to the industry."* Shit heads are not the norm, they're actually pretty rare. The problem is that, even though organizations tout the importance of the role, they broadly treat safety practitioners as shit heads. They regularly employ tactics that should only, if ever, be used when dealing with shit heads. Organizations often start from a place of distrust of the practitioner, distrust that is compounded by the practitioner frequently failing to meet the near-impossible expectations set forth by the organization. This mixture of mismatched and conflicting assumptions, coupled with many of the other beliefs about the safety practitioner that we have discussed in previous chapters, leads safety people to ultimately be "othered" by their organizations.

OTHERED

To view or treat (a person or group of people) as intrinsically different from and alien to oneself.

Merriam-Webster

Long after the majority of the workforce has left for the day, while managers and leaders are already home enjoying time with their families, you'll frequently find the safety professionals car still parked at work. Often, it is the vehicle that is the closest to the entrance, only being beaten out by the crews that worked night shift. Safety professionals find themselves working sunup to sundown; there's too much to do and not enough time or hands. Throughout the course of their days, they are barred from their offices, they must touch each and every person and must observe the majority of work. They must remain ever ready to swoop in and stop risky behaviors or respond to an event. While journeying through the field, they are expected to always be the shining example of safety, an example that everyone can and should aspire to be. Even in their personal lives, they often find the same expectation. They must demonstrate that they always *"carry safety home with them!"* They are prompted to share their (totally true) stories about how they always wear a full-face respirator, chaps, gloves, and steel-toed boots every time they mow the lawn. They must never admit that they too have sent a text message while driving or that they have ever taken any risks. They are far too "safe" to ever ride a bike without a helmet or partake in any risky recreational activity. The safety practitioner is to be barely human.

The safety professional is "othered;" they are viewed and treated as intrinsically different. They are "othered" by the views held about them, that they are always to be this non-human infallible beacon of safety. They are "othered" in treatment by the application of unreasonable and unachievable expectations, demands that are rarely placed on anyone else. Safety practitioners are often excluded from company initiatives and "feel good" events, they are far too busy in the field and the work of safety is far too important for such things. They find themselves to be an outsider in the organization; the person that showed up but wasn't invited to the party. Organizations are left in a state of shock and disbelief when the practitioner finally calls it quits and moves on to a competitor. They call them ungrateful and treat them as a traitor when they can take no more and flee for greener pastures. How can we move the safety practitioner from this "othered" status? How can we better care for safety professionals? How can we retain our talent?

Let's start with a simple concept, one that seems to be foreign to many organizations: *Treat people how you would like to be treated.* A gigantic first step in de-othering safety practitioners is to simply treat them as fellow humans, as normal people. While this seems simple on the surface, our assumptions and beliefs are ever present to hold us back. We like the idea of *safety being everywhere and always*, it fits our overly simplistic beliefs that the safety professional should be ever present to acutely prevent bad things from happening. Organizations enjoy having *a highly visible point of both action and blame* in the role of the practitioner. Thrusting blame for negative events upon the safety professional feels good, and by assigning them more and more *Safety! Fix It!* tasks, it feels as if we are really fixing problems. But we are not, as good as it feels to the organization, we are only harming the well-being of the practitioner while pretending that we are making an impact on worker safety. By acknowledging the humanity of the practitioner, we are challenging these beliefs and admitting that they have limitations to what they can influence and what they can do. From this acknowledgement and acceptance can begin to grow a more realistic, fulfilling, and impactful role for the safety practitioner.

As an organization, if you'd like to know what the professional safety practitioner should be investing their time into, ask the safety practitioner. If you'd like to create a better working environment for your safety folks, ask them! Craft an environment in which honesty is possible, approach with trust of the practitioner, and ask them. Rather than leaning into the commonly held beliefs and assumptions about what safety practitioners should be doing and how they should be treated, ask one that works within your organization. Their answers might be hard to stomach sometimes, but you're guaranteed to learn something. While I do not plan on rehashing the chapter *"what should we be doing"* or going back over what we have already touched on, I do want to leave you with a familiar question, one that I hope oozes from this book: What is the *Shit That Really Matters*? You'll find the answers to job expectations, crafting an impactful role, professional treatment, the care and feeding of safety professionals, and much, much more as you embark on this thought exploration of *What Shit Really Matters?*

FYI

Safety People Are People, Too!

With our newly discovered truth, one that says *safety people are people too*, we can begin to explore the care and feeding of safety practitioners. How can we better care for our safety folks? How can we curate an experience that is conducive to attracting and retaining safety professionals? What do safety practitioners desire in a workplace? Safety people are people, too. They seek out many, if not all, of the same human desires most look for in a workplace. They seek a sense of belonging, they want their input to be heard, they want to trust and to be trusted, they want dignity and respect, they want to make an impact, they want to do a good job, they need community, they crave friendship, they need a *"job well done"* from time to time, and they want their compensation to match their level of effort. Safety practitioners crave work/life balance, they need time away to relax, they want advancement opportunities, they want education and professional development opportunities, and ultimately, they desire to be treated the same as everyone else. They move towards these basic humans needs, and they move away from painful negatives. They run full speed, hardhat in hand, away from beatings, from being othered, and from *Safety! Fix It!* To better care for professional safety practitioners, quit treating them as non-human, and start treating them like normal people. Treat them how you would like to be treated, acknowledge that they have basics needs that must be fulfilled, and work towards creating that existence.

"You get back what you put out into the world," this is especially true as it relates to seeking out and hiring safety practitioners. Unfortunately, we often "put out" much of what we've talked about. Our job postings ooze safety guru, they wreak of "*Safety! Fix it,*" they scream the old definition of the safety practitioner, they stink of all work and no life, and they paint a clear portrait of a professional existence the safety practitioner hopes to avoid. We are left shocked by the lack of quality applicants; we only manage to lure the newest of new oblivious few into our trap. If you hope to attract amazing and talented safety practitioners, start by crafting the work life experience that we have already discussed, one that is based on their innate human needs. Build your job descriptions and postings around our new definition for the safety professional and mean it! Focus on the practitioner being a communicator, facilitator, team member, team builder, and curious innovator. Rather than asking for a compliance cop, ask for a curious person with an obsession for learning about work. Ask for an evangelist of learning and organizational betterment, rather than a fortune teller and fixer. Build a better environment, and meaningfully ask and you shall receive.

Safety practitioners have been mistreated, othered, and abused for far too long. Unrealistic and extreme job demands, ones that have doomed the safety professional to a life of failure, have taken their toll. The expectations that they be a fixer, a near-instant responder, a guru, a safety saint, and an easy button for the organization have left safety folks beaten and bloody. The othering of safety practitioners, especially as it relates to work/life balance, long unpaid hours, and differential treatment, has kicked many safety professional's asses, and left them burned out. Safety people are people too, and it's about time that we started treating them as such. To better care for professional safety practitioners, we must start by acknowledging and embracing their humanity. We must then work to craft a work experience that is based off of their needs as humans, ever trying to treat safety people how we would like to be treated.

The Mob

The world of occupational safety and health is in turmoil; what an odd and polarizing time we currently live in. When I first started my journey in the professional practice of safety, I could have never imagined that safety would become such a touchy subject. Our work worlds seem to mirror the worlds that exist beside them, this deeply entrenched polarization is more the rule than the exception in our current lives. We find ourselves in a world of safety this, safety that, and (*insert some new fancy, slick, and marketable safety related name here*) that we can all argue about online. We have broken ourselves up into pseudo-political factions, abandoning free-thought and reason, in exchange for cookie-cutter platforms constructed for us by others. We bolt these ideologies onto ourselves like suits of armor, we raise the flag of current chosen team, and we go to war over our beliefs. We have labeled and boxed ourselves into our respective corners, we have set up camps, we have drawn a line in the sand, and we are more than willing to duke it out to the death for our chosen side.

This creation of safety factions has left much to be desired. We no longer have schools of thought around safety; respected groups with opposing or differing views on how to best approach the complex problems that we face seem to have become a thing of the past. Open conversations, dissent, diverse thought, and mutual professional respect, seem unwelcome in our current safety ecosystem. Rather, we have rival safety gangs roaming the digital streets of social media hoping to start a turf war with each other, and practically anyone else for that matter. If you don't believe me, I double-dare you to log onto *LinkedIn* and post something like, "*HOP Sucks!*" or "*Behavior-Based Safety is stupid!*" and see what happens. If you what to get really dirty and freaky, mention something about *ZERO Injuries*! No matter your preferred poison, you will soon be swarmed by an unruly mob, some coming to your aid and others prepared to deliver a digital cement milkshake to your face. Look, being a bit of a scrappy provocateur myself, I like a good fight as much as the next person. But we seem to be fighting simply for the sake of fighting. On our best days we seem to be more interested in fighting over what we label ourselves, our little groups, and our ideologies, more than we are interested in fighting to make things better. On our worst days, we seem to have no idea why we are fighting to begin with, but the fight continues. We have devolved into a feud-like state resembling that of the *Hatfield's and McCoy's*, and why it started, along with the reasoning for continuing our safety feud, seems to be just as obscure.

Before we move forward, I want to state my unabridged support for real and meaningful conversations. Open dialogue, even dialogue consisting of opinions that we dislike, matters. This meaningless feuding and fighting though, is only noise. This screeching, screaming, and open-air online feuding prevents or discourages meaningful conversations, dissent, and debate. Rather than being dismissive or argumentative, we need to seek to understand each other's views and ideologies. We need less "seagull" interactions, interactions in which one person flies by and shits all over someone or their beliefs and then flies off, and they should be replaced with meaningful conversations and debate. We need to place less value on *"Ha! I was right and you were wrong!"* and place more value on *"Ha! We discovered the truth, together!"* We need to try harder to understand one another, and the ideas that we each hold. Even if the ideas aren't great, even if they are horrible, we need to be able to talk about them in a constructive fashion. Bad ideas fester in darkness; bad ideas are often born and thrive in a vacuum. So, even the worst of ideas are better suited to reside in the light of day. In the daylight they can be openly challenged by the many; in the darkness they can be perpetuated by the few.

  We need to talk more and fight less! The conversations that we have, the dialogues that we create, they shape the worlds in which we reside. So, this is in no way a war cry against dissent, debate, or even good old-fashioned arguments. It's actually quite the opposite. What I am really saying is this: *We've been shitting where we sleep for far too long.* While it might be fun and sometimes warranted, "shitting" all over our peers, is still shitting on our own! And it is still shitting in our own bed! Not only do we continually shit the bed, but we roll around in it, basking in how victorious we were over that other "dumb" safety practitioner. We seem to prefer being "right," gobbling up any practitioner that dare disagrees with us along the way, to seeking out truth and trying to actually do *Safety Better*. Here's the point, arguing simply because you're on "team this" and I'm on "team that," is only arguing in pursuit of <u>being right</u>, not actually seeking out <u>what is right</u>. We hide behind this notion that our group of believers holds the truth, not only "the" truth, but that our ideas hold the moral high ground as well. We hold in our hearts that those with different views are not only dead wrong, but morally inferior. This drives us to "eat our own" rather than seeking to understand why these particular views make sense to this particular person, in this particular situation.

We make some pretty bold assumptions that only our team has the right answers, that only our ideologies are true, that there can only be a singular truth, and that those who do not fit neatly into our chosen belief system are somehow less valuable. I hate to break this news, but I do not hold all of the answers. Hell, I even get it wrong a lot! Worse yet, you do not hold all of the answers either, and you get it wrong a lot, too! Even worse, your chosen ideology isn't always right, even if it is super new and slick, or super old and covered in cobwebs. To make matters much, much more complex, more than one thing can be true at once. Do you hear those loud clacking noises and primal grunts and screeches? The social media mob is already forming to burn us at the stake for daring to have such a blasphemous conversation!

The point is this, we need each other. We need each other's support, thought, creativity, know-how, knowledge, and wisdom. We each hold bits of truth and knowledge, like those bad ass superhero rings we have all seen in movies or comics, they work best when combined. Each is powerful on their own, but when combined we rise to another level completely. When we come together, we can actually create something pretty cool. But in order to use it, we have to actually come together as professionals, leaning into our commonalities and embracing the diverse thoughts and opinions we each hold. We need to understand that most of us are on the same mission, *to make the world a better place to work*, even if we disagree on how to make that happen. Rather than bickering so much about the parts that differ, isn't it about time that we lean into the objective truths that bind us together? Should we not lean harder into growing a diverse community of practice, rather than seeking to beat each other down? Rather than "burning it all to ground" only to be rebuilt, and potentially rebuilt worse, it seems better to come together and objectively examine where we have been, where we are at, where we hope to go, and how we get there together.

As much as I often favor anarchy, I've never really understood this desire to "burn it all down" as it relates to safety. Our systems, structures, and hierarchies have come from somewhere, they were built upon years of generational knowledge, they may be flawed, but they are not valueless. We shouldn't be so thirsty for change, so frustrated with our current state, that we're willing to set a match to the world just to watch it burn. Our job is to make these systems better, sometimes that means burning parts of them, or burning the total system entirely to scorched earth, but not always. The right amount of fire can be cleansing, but too much can cause more harm than good. These structures and frameworks are not always the enemy, and we should, at the very least, seek to understand them before we torch them. Oops, my minarchist leanings are showing. We seem to struggle with where we have been, and with where we are going. Hell, half of the time we do not appear to even know where we are at in the present. We cannot seem to reconcile our history and our present state with our future. We find ourselves so obsessed with what's next, that we lose sight of what's happening in the present. We get so tied to the past, that we can't seem to imagine anything beyond it. Either way, we seem to misjudge or undervalue the power of our history and the importance it has in shaping where we go next.

PRESENTISM:
an attitude toward the past dominated by present-day attitudes and experiences

Merriam-Webster

As we peer back into the history of safety and health, we seem apply a strong dose of presentism, we judge past iterations of worker safety by today's standards. I'm guilty, you're guilty, we are all guilty. We take all of the vital information that we now hold, all of the lessons that we have learned along the way, todays moral standards, and all of the collective wisdom that we have attained, and we use it as weapon to undercut the history that has led us to where we are currently at. We debunk, we disprove, and we banish these now dated ideologies from our purview, a lot of which is a good thing, but we do not just stop there. We tend to banish the often-well-meaning practitioners that promote these views, and we often *"toss the baby out with the bath water,"* forgetting our history, losing vital learnings, eroding the building blocks on which we stand, and knowing less than when we started. Is it any wonder that we find ourselves at each other's throats so regularly?

Inversely, we tend to sugar-coat the bits that we favor, have history with, or have strong feelings for. We fluff up, puff up, and look past the gnarly bits of the historical approaches to worker safety that we have a certain personal affinity for, or attraction to. We do much of the same as it relates to where we are going as well, we sugar-coat as we examine more modern views of worker safety and safety systems at work. We will take some ideology like *Behavior-Based Safety (BBS)* or *Human and Organizational Performance (HOP)* and we will choose to only see what makes it great, all the while we are careful not to admit any potential pitfalls or downsides that might be contained within its particular set of beliefs. We choose to see and promote what is obviously positive, avoid or dismiss the things that are obvious negatives, and view honest critique as desecration of our sacred ideologies.

This sugar-coating is a common phenomenon, beyond our work worlds we see it manifest in our accounts of history, in our religious leanings, and within many other bits of our lives. We sugar-coat, we cherry-pick, and we seek truth, just not the whole truth unless it aligns with our preconceived notions or assumptions. As we skim our religious texts, we are quick to point out the buckets full of love and forgiveness while we swiftly avoid all of that pesky wrath, fire, and brimstone stuff. We look back on the history and culture of the antebellum American south with almost child-like wonder and amazement, and we attempt to look past the vile stain that was slavery. Worse yet and as mentioned, we view it all through the flawed lens of presentism that seems to entrap us within this "all or nothing" thought prison.

Now, keeping in mind that certain "rights and wrongs" have almost always been right or wrong, and they will continue to be, whether we are discussing the American south or religion, traditional safety or *safety differently*, *Behavior-Based Safety* or *Human and Organizational Performance,* or practically anything else… our mindsets leave us insisting that that they are either completely good or completely evil, period. We have left no room for reason, nuance, or objectivity. This "all or nothing" mindset has removed our ability to learn from things outside of our ideological vacuum. We have placed ourselves in a position in which a non-believer has no ability to learn from biblical wisdom, in which science cannot learn from myth, where newer views of worker safety cannot learn from more traditional approaches or vice versa, and a position in which good and bad are viewed as always opposing rather than often intertwined. What an illogical, boring, and unproductive place to find ourselves.

We need community, not rivalries. We need conversations, not *LinkedIn* hit squads. Our desires to cling to our particular flavor of safety ideology above all else, is destroying valuable community in the professional practice of safety, a community that we are all in desperate need of as we work within this challenging profession. No particular safety ideology is the root of all evil, nor is it all that plagues us as organizations or as safety practitioners. Screaming at each other about how great "safety this" is and how dumb "safety that" is, is only harming all of our efforts to *make the world a better place to work* and driving us farther away from each other. If something is wrong, say it! But dissent, debate, explain, teach, rather than shit on, exile, disavow, and disembowel. People are not the enemy of better, crystalized ideologies are the real enemy of better. They are tearing us apart and preventing us from growing a viable and diverse community of practice. Crystalized ideologies, ideologies that we cling to so firmly that they prevent us from ever being wrong, from seeing beyond them, from learning, growing, or evolving, are as close to a defined source of evil you will find within this book. We need to maintain open minds, open hearts, and we need to be open to learning from each other. Rather than fighting for our respective teams or destroying each other to prove how right and righteous we and our held beliefs are, we need to come together to seek out better and to seek out truth. In case that offends you, I'll leave you with the words of a wise Texas philosopher, Matthew McConaughey, "*I believe the truth is only offensive when we're lying.*" Remember, *the only people mad when you speak the truth are those that are living a lie.*

Never Stay in Your Lane!

Comfort is a hell of a thing. Its enchanting call leads people to avoid trying new things, keeps them in toxic relationships, makes folks quickly abandon their *New Years Resolutions*, causes them to miss amazing and life changing opportunities, and its pursuit ultimately leads to an existence of mediocrity. It drives organizations, industries, and professionals down the path of seeking "good enough" rather than what's best, it causes them to double down on what they already know rather than trying to discover better ways to do things, and it pushes them to "always stick to the plan," even if the plan doesn't work. We've all heard a few of these, or similar, calls to stay in our comfort zones and pursue adequacy while working within the safety profession:

> *Stay in your lane!*
>
> *Stay in your wheelhouse.*
>
> *Let's not reinvent the wheel.*

What are these common corporate phrases really saying? What do they really mean? Often, they're explained away as positive monikers touting the benefits of focus. We'll hear such things as, "Stay in your lane, it's the best use of your time and energy," or "Let's not reinvent the wheel, we already have that solved!" While there is an ounce of truth when used in these typical fashions, there is also a pound of bull shit. What these everyday phrases, and the beliefs that often hide behind them, are actually saying is more along the lines of, "mind your damn business!" And "that's not your job!" We might as well be saying:

> *We don't need any of your ideas...*
>
> *Your input can't possibly be valid...*
>
> *Safe mediocrity is better than risky innovation!*

"Staying in your lane," while seemingly safe and beneficial to focus, causes a lot of harm. The negative results of avoiding diverse and varied input, experimenting, and stepping from comfort into risk can kill, or at the very least cause us to miss out on, innovation, creativity, engagement opportunities, and learning. But our old friend comfort is ever present to tell us otherwise. Comfort, both within ourselves as professionals and the organizations in which we work, is always there whispering sweet seductive nothings into our ears. Comfort will tell us:

"Are you sure you want to try that new approach? It's never been done that way before!"

"Your peers, the ones that you invest so much time and money into benchmarking, don't do it that way. You'd better stick with the herd!"

"You know, even if it might be a better way to do that, risk is scary! You should stick to what you know, at least you know that it's safe!"

And on, and on....

Investing in comfort feels like we are eliminating risk; comfort gives us a feeling of safety and security. It leads us to believe that by not "reinventing the wheel," we are avoiding the perils brought about by change and growth. But in fact, we are only exchanging one set of problems for another, more troublesome set of problems. Let's boil this down to a simple example: *We are getting better, or we are getting worse*, there is no such thing as "maintaining," and "the middle" doesn't exist. Now, I understand that "the middle" does in fact exist. But, for the sake of our discussion let's make a couple broad assumptions about it, *a.) the middle kills innovation and progress, b.) the perceived state of stasis or "maintaining" is an illusion. We are in a constant state of degradation.* Simply put, we are either getting better or we are getting worse. When we believe that we are maintaining, we are in fact getting worse.

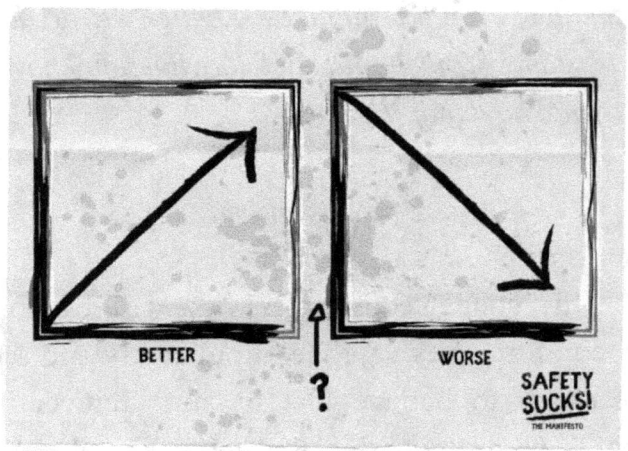

So, as we focus all of our personal and organizational powers into maintaining the status quo, we are not "maintaining" at all. Comfort leads us to believe that by staying the same, by avoiding experimentation and adaptation, and by leaning into safe mediocrity over risky innovation, we are somehow making the safer and wiser choice. But in all actuality, our constant state of degradation is resulting in compounding risk that will eventually come to a head. When the status quo eventually fails, it will leave us dumbfounded. We will find ourselves asking the same familiar questions, *"how did we not know?"* *"How did we not see?"* *"How did we fail to predict this?"* Worse yet, we will follow up these horrible questions with just as horrible answers like, *"We just need to do the same thing harder!"* or *"What does everyone else do for this? Ok, let's do that!"* This illusion, the one that leads us to believe that there is safety in stasis, is simple and comforting. It aligns with our wishes of how we want the world to be, rather than the reality of how it actually is. This idea that, if we can finally just plug all of the holes and place ourselves in a holding pattern then all will be well, should frighten us as companies and as safety practitioners. We are getting better, or we are getting worse, there is not an in between.

Why do organizations find themselves clinging to the things that they know are ineffective? Why do we hold fast to these ideas and approaches even when they are proven to be harmful?

> *Stasis seems safe and feels comfortable*
>
> *Change is hard and scary!*

There is an old saying, one that is attributed to various people, that says, *"We change when the pain of staying the same becomes greater than the pain of changing."* Unfortunately, that pain as it relates to companies and safety in particular, is usually the result of some catastrophic or fatal event. Sometimes, it only comes about as the result of several of these types of events strung together over a period of time. Stasis and mediocrity seem great until we start killing people or blowing things up. With this newfound source of suffering, we seek to quickly eliminate it. Surprisingly, but typically, we still avoid the pain of embracing risky innovation. Our innate desire for comfortable stasis over discomfortable progress, causes us to lean that much harder into doing the same things we've always done. Rather than seeking learning and growth; refocusing and increased effort on "the way we have always done things" seems like the safest and easiest bet. Comfortable mediocrity is powerful, so powerful that even when killing people, we run into its open arms. *The pain of killing or maiming people has not become greater than the pain of changing.* Now, it eventually will, but how long will that take for your organization? As industries, what is the magic body count that finally justifies a break from the norm and a turn into more creative approaches? Additionally, if occupational fatalities are not enough motivation to embrace innovation, can we honestly expect less important areas of the business to embrace risky progress and change? Betterment seems far away when our vivacious desires for lockstep uniformity and mediocrity appear unwavering.

    Organizations desire progress without change, they desire innovation without risk, and they crave creativity without adaptation or unpredictability, none of which is possible. A high value is placed upon the maintenance of the status quo. There is a firmly held belief that safety and comfort can be grown by doing business as usual, just harder and with more focus. This belief, although working counter to innovation and progress, is pushed down the line to leaders, frontline employees, and safety professionals alike. Its sentiment manifests in the ways that we react to surprises, new ideas, growth opportunities, and challenges. It creeps into the words that we use when we are responding to our employees that are proponents for betterment, agents of change, or curious innovators. When employees take on the interpersonal risk of seeking to affect change within our organizations, we are quick to respond with the tried and true, *"just stay in your lane,"* *"stay in your wheelhouse,"* *"that's not any of your concern,"* or *"let's not reinvent the wheel!"*

Innovation, experimentation, and change are all accompanied by a healthy dose of risk. As with all that we seek to do in the course of our lives, with progress comes the potential for failure. But, as Bill Gates once wisely said, *"Success is a lousy teacher. It seduces smart people into thinking they can't lose."* Elon Musk, the founder of such companies as *Tesla*, *SpaceX*, and the *Boring Company* once said in reference to failure, *"Failure is an option here. If things are not failing, you are not innovating."* Elon Musk knows a thing or two about both innovation and failure. Earlier in his life he was rejected by *Netscape* and ousted from *ZIP2*, after founding *PayPal* it was voted *"the worst business idea of 1999"* before becoming the massively successful business that it is today, *Tesla* failed in more ways than imaginable before they found success, *SpaceX* blew up rocket after rocket before becoming viable, and on, and on. The lesson of Elon Musk is one of embracing the relationship that exists between innovation and risk. It is a lesson of constantly pursuing "better," even if that pursuit comes with the potential for failure. It is a lesson, even though sometimes terrifying, to lean into risky innovation rather than safe mediocrity. It is a lesson that safety practitioners, and the organizations that they serve, could stand to learn.

As safety practitioners, we desire comfortable mediocrity just as much as the companies that employ us. We tell ourselves, and each other, many of the same things. *"We just need to stick to our lane," "I know safety and health, I don't need to put my nose in other areas of the business where it does not belong," "We've done safety this way for a really long time, we shouldn't reinvent the wheel now,"* and much, much more. Safety innovation frightens us, it paralyzes us. The old "tried and true" approaches that we know all too well, are comforting and appear to be less risky than new and less proven methods. So, we do exactly that, we stay in our lanes, we keep to our little wheelhouses, we avoid reinventing the wheel, and innovation and creativity within our profession withers and dies.

As safety practitioners, we align ourselves with the strongly held belief that our companies and industries hold that innovation and creativity, within the world of safety, is far too risky to pursue. We join in with the "there is no acceptable level of risk" choir, singing about the perils of doing safety things differently, and we preach from our risk adverse pulpit that, *"if we experiment or innovate in safety and health, then we are gambling with our workers lives!"* As great as it feels, and as seductively simple as it seems, nothing could be farther from the truth. In fact, a healthy argument can be made that by forcing mediocrity, and by dissuading innovation, that we are taking a far greater gamble than by innovating and creating betterment. There is a long-proven track record with our stagnant, stale, and starchy "tried and proven" ways. One that demonstrates that we are remarkably good at consistently maiming and killing people at a steady rate, year-over-year. Our "tried and true" approach is finely tuned to yield the results that we so consistently see. Yet, we think that if we just do it harder and faster, something will change. If we can only do mediocre, or ride "the middle" better, all while avoiding risky innovation, then surely things will get better. We try and try, but they never do. As Albert Einstein so famously said, *"the definition of insanity is doing the same thing over and over and expecting different results."* Our deeply rooted desires for comfort, predictability, stasis, and uniformity, along with our crippling fears of innovation, change, creativity, and unpredictability, has resulted in *safety insanity*. An insane state that we cling to like a warm security blanket, even if it costs our employees their lives.

As safety practitioners and companies, we can no longer afford to ride "the middle." We must shift our thinking from managers of the status quo to fearless, yet risk aware, innovators and creators. Ultimately, we must choose to seek out risky innovation over safe mediocrity. Staying the same by avoiding innovative and new approaches, even when we do those same things harder than we ever have before, will only leave us with the same results. We will tell ourselves that we are avoiding risk and that, "we are using tried and proven ways to keep our people safe!" But all the while, our systems are degrading, and our fear of innovation is crippling our businesses and our people. Change is one of the few things in life that is guaranteed. We can embrace this fact my marching forward into innovation and betterment, in essence, creating the change that we want to see. Or we can cling to our "zero risk!" security blankie and have change forced onto us, never knowing what it is, where it will come from next, and completely lacking any ability to influence it. In a chaotic and risk filled world, a world in which change is guaranteed, innovation and adaptability is king. We must embrace and always pursue innovation, even with failure being a glaring option, if we want to make things better.

Now, there is also a lot to be said about "not staying in your lane" as it relates to individual personal and professional growth. With these firmly held beliefs, beliefs that we have already touched upon, it's easy for safety practitioners to miss out on a lot of learning and growth opportunities. It's easy for us to assume that the best way for us to get really, really good at safety, is to invest all of our time and energy into getting really good at safety. That sounds pretty good if taken at face value, but it's a big world out there and there are many places from which we can derive knowledge. As a safety professional, learning about safety and health is a fairly good idea. But it is a far cry from the only thing that you'll need to invest time into learning about. In fact, there are probably a few areas that are even more important than health and safety proper we should be learning from. Blasphemous, I know! Look, I'm not arguing against a solid understanding of the basics, nor am I discrediting health and safety education. What I'm saying is, those alone will never be enough.

If we "stay in our lane," and health and safety is all that we seek to know, are we really curating the unique and needed skillset that we require to be successful as professional safety practitioners? I'll call your attention once again to some elements of the better-defined role of the modern safety practitioner:

THE ESSENCE OF THE MODERN SAFETY PRACTITIONER

THE SAFETY PRACTITIONER SHOULD BE A.......
FACILITATOR
COMMUNICATOR
TEAM MEMBER
TEAM BUILDER
CURIOUS INNOVATOR
EVANGELIST FOR BETTERMENT

If we "stay in our wheelhouse," can we ever hope to grow into that new and better-defined role of the safety practitioner? We will surely miss the mark if we contain ourselves to the rigid confines of safety and health proper. We must stick our noses into various subject matters, we must stay curious about other professions and industries, we must approach this buffet of learning with a healthy appetite and a plate in each hand. Rather than calling out the numerous areas in which we as safety professionals should be seeking out knowledge and "know-how," I'll bring you back to a question that we have already asked several times in this book, *what's the shit that really matters?* What really, really matters? Based off of our new and better definition of the safety practitioner, what subjects should we be exploring? One thing is for certain, although valuable and needed, the study of health and safety alone will never be enough. If our desire is to create the most impactful role for the safety practitioner, we must never stay in our lane.

As organizations and as safety practitioners, we double down on what we know, and we lean into comfort. We find solace in the status quo. Comfortable is exactly that, it's comforting. We hang our hats on the old "tried and true" methods, trying to do them harder and faster when they ultimately let us down. Within our little worlds, we work feverishly to maintain some form of orderly stasis, all the while avoiding risky innovation and unpredictability. Our sense of comfort and security is simply an illusion, a teddy that we cling to in lieu of innovation and potential failure. Along with stasis comes stagnation and degradation, despite our best efforts to stay the same, we are only getting worse. On the shoulders of innovation rides both failure and progress, we need both to create betterment. We must learn, try, fail, learn, and repeat. Within this cycle of innovation, not within the good old "tried and true," we will discover the answers that we so desperately seek.

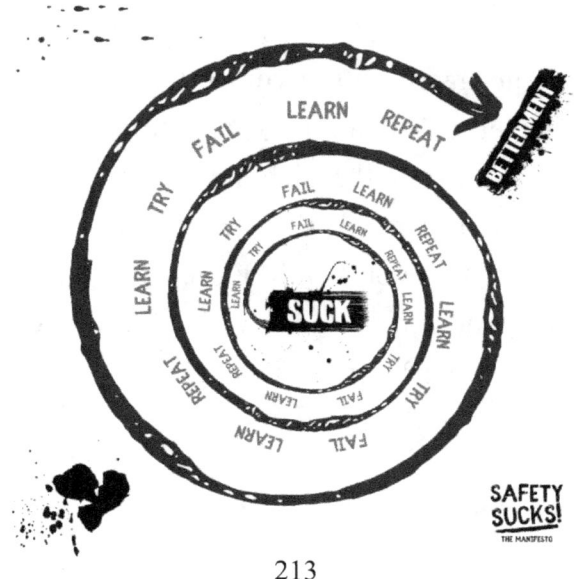

If our hope is to put an end to occupational maiming's and fatalities, if we hope to make work suck less and to better worker's lives across the board, if we wish to create an impactful role for ourselves as professional safety practitioners, then we must move away from safe mediocrity and embrace risky innovation and learning. If we hope to have the "know-how" and knowledge bank to pursue such worthwhile endeavors, we can no longer allow ourselves or others to be limited by harmful idioms such as "stay in your lane!" Fuck staying in your lane! Swerve all over the damn highway with your head out of the window screaming to the top of your lungs about doing things better! Let yourself go wherever learning takes you! You curious innovator, you! Don't fall victim to comfortable stasis, within yourself or the organizations that you serve. Lean into discomfort and embrace innovation. From the rich fertile grounds of failed innovative attempts, grows a crop of progress and betterment. But you'll never see your harvest if you "stay in your lane" and build a career of playing maintenance man for the status quo.

The Wrap Up

We face a challenge; an extremely daunting task is ahead of us. But every journey, no matter how long and perilous, starts by taking the first step. Before we can take that step, as a profession and as organizations, we must decide that it is something that we wish to pursue. With every adventure comes the opportunity for great risk and reward, but through our explorations we have the opportunity to discover ways of doing *safety better*. Edith Widder, an American oceanographer, marine biologist, and veteran of over 250 dives in the *Johnson Sea Link* submersibles once stated, "*exploration is the engine that drives innovation…*" This journey, our risky and rewarding exploration of all thing's *safety better*, is a worthwhile pursuit if we wish to better the safety of our workplaces and create a more impactful role for safety practitioners.

The pain of staying the same, is a pain that we know all too well. We've talked quite a bit about our desires to stay the same and still see different operational results. Are we finally ready? Are we prepared to let go of our current state of *safety insanity*? *Has the pain of staying the same finally become greater than the pain of changing*? Our inherently dangerous businesses kill and maim consistently, year after year. This pain-point alone should be enough to send us off adventuring into the new world, and it should be enough for us to burn our ships when we arrive. But unfortunately, our deeply rooted assumptions about safety in practice and our desire for a comfortable change-free and danger-free world, leads us to keep our vessels moored in safe and well-known harbors. Our simplistic beliefs about safety are comforting, the approaches that stem from them appear easy and "commonsense," we are enchanted by this idea of a "safe" la-la-land where failure never occurs, a world that we can create if we eventually make people care enough or pay enough attention, and this prevents us from ever undertaking an exploration in doing *safety better* to begin with.

No matter how hard we try, the dangers of work will be ever present. We have fooled ourselves into believing that, through repetitious mediocrity, we will finally eliminate danger and risk from our work worlds. We have clung to this idea that by eradicating bumps and scratches from our worlds, that we will finally stop killing people in confined spaces. We have doubled down on this notion that when we finally reach peak caring and awareness, people will stop dying from electrical contact. We have crafted this utopian end-state image in our heads of this world in which the possibility of suffering, harm, or injury does not exist. We have been fooled into thinking that we create this idyllic world through preventing every possible bad thing from ever happening, and if it does happen, it shouldn't have! While we invest all our time and resources into a sole focus on prevention, while we seek out "safe" statis and mediocrity, while we seek to influence our complex and dangerous worlds through simplification, while we tell our people to care more and follow rules harder, while we count days "since someone last messed up" and throw pizza parties for long streaks of *OSHA* recordable-free performance, our people are suffering and regularly dying. I'll ask once more, are we finally ready? Are we ready to explore new worlds in search of safety innovation? Are we ready to embrace risky innovation in pursuit of doing *safety better*? Are we ready to get comfortable with being uncomfortable? *Has the pain of staying the same finally become greater than the pain of changing?* We must be ready to change, or we should be well-prepared for a long and frustrating existence of sucking at safety.

For most of us, especially if you have made it this far in this book without burning it, the case for innovative change is clear. Investing our brief amount of time together into further explaining why and how this change is extremely beneficial for organizations, safety practitioners, and employees, is probably the least effective use of our time. Again, assuming that this book has not made its way to the fire pile due to its blasphemous nature, I'm taking a guess that you "get it." A better use of our time, in keeping with our focus on the *Shit That Really Matters*, lets focus on pulling this all together in practice. Let's revisit a familiar and vital place, our first focus area for any meaningful and lasting change, a shift in underlying assumptions.

Our underlying assumptions lead to the creation of our values, which leads to how we think about things, the actions that we take, our approach to problems, how we approach betterment, and other exposed bits that connect back to our deeply held beliefs. For many of us, we refer to this phenomenon as "culture." Edgar Schein, former MIT professor and author of such works as *Humble Inquiry* and the *Corporate Culture Survival Guide* said, defining culture as, *"a pattern of shared basic assumptions invented, discovered or developed by a given group as it learns to cope with its problems of external adaptation and internal integration that have worked well enough to be considered valid and therefore, to be taught to new members as the correct way to perceive, think and feel in relation to those problems."* We have spent a large amount of time picking at and pulling apart some of our deeply rooted assumptions about worker safety and the role of the professional safety practitioner, but these insightful words highlight another challenge to influencing *"the ways we've always done things."* Simply put, they have worked. At the very least, they have worked "good enough."

Over the years, our drive to prevent bad things from happening has indeed worked. A quick examination of data around occupational injuries and fatalities demonstrate a sharp decline in the total number of significant injuries and deaths over time. A great approach to highlight, one that has aided progress in safety, is prevention. Prevention is a great thing, it has led to numerous better outcomes, both in our personal and work lives. Automobile technology is a great example of neat and useful prevention tools. We have back up sensors, cameras, blind-spot detection systems, and a variety of other tools and gadgets that are designed to aid you in not wrecking your car. Similarly, in the world of worker safety, we have created a long list of gadgets and processes that do the same. Guess what? Prevention works. But let's add in an even better question, let's throw a monkey wrench into the gears of prevention, for how long? How long until: 1) *our preventative strategies breakdown and fail*, 2) *they reach their peak effectiveness and are no longer viable*, and 3) *the overreliance on prevention and doing prevention harder creates more harm than it does cure.*

Within the world of safety, we draw a hardline in the sand that says, failure must never occur. Based off of our over reliance on prevention, when bad stuff inevitably happens, we find ourselves asking the same old tired question, "how did we fail to prevent?" We consistently stay on the straight and narrow path of prevention because we assume that a) *every incident is preventable* and b) *by focusing on preventing little events, we prevent larger and more catastrophic events.* This crafts our current assumption about what safety is, and how we seek to influence it. It shapes our views that safety is the absence of accidents and we influence safety by preventing accidents. We draw the conclusion that we put an end to occupational maiming's and fatalities by preventing the "little things" that we inaccurately believe lead to them. With all of this in hand, we never think beyond prevention. Our assumptions that all failure is preventable and that it must never occur, confines us to prevention as our sole approach.

Diving back into the world of automobile safety, our friends in that industry got really, really great at prevention. But no matter how hard they focused on keeping cars from wrecking, what happened? Cars continued to wreck; people continued to die. A shift in assumptions was needed to bring about the change that they so desired; prevention alone would never be enough. While understanding that prevention will always be something to focus on, car manufactures expanded their thoughts about failure. Rather than saying *"our goal is to prevent cars from ever wrecking"* by assuming that all failure can prevented, they shifted towards better underlying beliefs. Now, understanding that failure is always an option, that assumption sounds more like this, *"No matter how great we are at prevention, car wrecks will happen. When they do, let's make sure people survive them."* From this new and better assumption has grown massive improvements in airbag, crumple-zone, and seatbelt technology, along with numerous other innovations. Some car manufactures have even created "fatality-free" cars, cars that have gone years and years without an occupant fatality.

    All of this long-winded discussion is to leave you with a clear focus area to begin bettering safety, a targeting of the underlying assumptions and beliefs that impact safety. We should take a page out of the car industries book; we should shift towards better underlying assumptions about safety. Rather than saying *"our goal is to make sure no one ever gets hurt at work,"* we should be saying *"no matter how great we are at prevention, accidents will happen. When they do, let's make sure people survive them and walk away from them as unscathed as possible."* This shift completely changes the game; we find ourselves planning for failure around the *Shit That Kills Us* and the *Shit That Really Matters*. While still understanding the importance of our preventative efforts, we begin to see that they will never be enough. We begin to understand that we do not have a prevention problem, in fact, we are already pretty great at prevention. I'll take a wild guess here and say that it's probably not normal for your organization to cut off arms, gouge out eyes, or kill people at work. If that statement sounds about right, odds are you're really good at prevention already. We do not have a prevention problem; we have a capacity problem. To begin to deal with this problem, let's take a stab at some of these better assumptions about failure:

SOME BETTER ASSUMPTIONS...

Failure will occur! Assuming for failure allows us to invest more time into recoverability and the ability to absorb failure with minimal outcome.

Our preventive strategies will always breakdown and fail at some point.

The little things, like bumps and scratches, are not predictive of more catastrophic events.

Ridding our organizations of the "little things," does not rid them of STKY.

With some better assumptions about failure clearly identified, let's revisit some assumptions about safety in general that we have already discussed:

OUR NEW ASSUMPTIONS ABOUT SAFETY...

Safety is the presence of positives, it is the presence of things that help create better outcomes.

All of the betterment that we seek is dependent upon the creation and maintenance of environments in which honesty is possible.

Learning is everything! It is the only tool that we have to do safety better.

Anything that harms, or works counter to learning or trust, is bad for our business and our people.

SAFETY SUCKS! THE MANIFESTO

Now, I want you to close your eyes for a moment and imagine what it would be like if your organization honestly believed those. What would our workplaces be like if we really believed that failure will always occur, that the ability to be honest matters, that things that hinder honesty or learning are horrible for us, that preventions is neat stuff but that it will ultimately fail, and that safety is the presence of things that generate better outcomes? How would things be different? What would change? If you reside within a more traditional organization, one that maintains many of the not-so-great assumptions we have previously discussed, there is one correct and glaring answer: everything. Absolutely everything would change. You now see the power of underlying assumptions and why, if we want true and meaningful change, we must work to better them.

Let's string all of this together into some clear picture of *safety better* in practice. I want to state that this is not meant to oversimplify the concepts that we have discussed in this book, its only meant to demonstrate how a better approach can be applied and its results.

SAFETY BETTER IN PRACTICE...

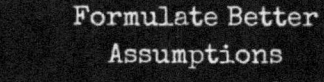

Formulate Better Assumptions

Safety is the presence of positives
Failure will occur
Learning is everything!
and more....

We Shift Our Approach

We start from a place of trust, rather than distrust

We do things with people, rather than to them

The role of the practitioner changes dramatically

Safety changes dramatically overall

All of the negative things that we experience while professionally practicing occupational safety and health, along with the problems we encounter with traditional approaches to worker safety, stem from our shared underlying beliefs about safety. If our desire is to create a better and more impactful role for safety practitioners, if we want to better worker safety, and our goal is to ultimately make safety suck less, then that is where we must begin. We must set aside our desires for fixing surface-level symptoms, and we must dive much, much deeper into the where the problems grow from. We must replace flawed assumptions with better ones, we must have open and honest conversations about these beliefs and where they come from, and we must embrace better approaches.

While I wish I could say something like, *"Just follow this super easy 5-step guide and world-class safety performance will be yours!"* or *"Just do X, Y, and Z, and the safety profession will no longer suck!"* I can't. Wouldn't it be nice if safety could be reduced down to some simple linear prescriptive method, one that guarantees consistent results, that we could all follow into the safety promised land? But that will never be the case. Our work worlds are complex, messy, and chaotic living organisms. With that being said, I think that it's important to again highlight how unique and complex your particular organization or workplace is by saying that this is not meant to be some prescription you blindly take or administer to your organization. Ultimately, I hope that you walk away with even more questions, and on a mission to learn. Not to learn from more and more safety theory, but with a thirst to go forth and learn from those that *Get Shit Done* within your organizations. Rather than prescribing to you "*the one right way*" to do safety, I hope that this book has influenced how you think about safety. I hope that it has provided you with a better lens in which to view safety, both from the vantage point of the safety practitioner, and the organization. I hope that it helps you craft and mold better assumptions about how we approach people, safety practitioners, and the safety of work. While I can't promise you some simple and easy fix to all that ails us in the world of safety, I can promise you this: If we start with better assumptions, if we approach all that we seek to accomplish from a place of trusting people, we do things with people rather than to people, we ask better questions, and we create environments in which honesty is possible, things can only get better and work will suck less overall.

About the Authors

Sam Goodman is a father, significant other, and a friend, first and above all else. He is also a creator, safety professional, and betterment evangelist. He is the host and producer of Really F**king Scary Stories, The HOP Nerd Podcast, and Hey Y'all with Sam Goodman. He is the founder of Pale Horse Media Co., HOP University, and co-founder of Project Visible. He is an accomplished author, speaker, consultant, and coach. Sam lives in Phoenix, Arizona with his partner Jerel and their amazing daughter Avery. Sam enjoys creating "bad a** things" and has made it his life's mission to "Make the World a Better Place to Work!"

For more about Sam, visit www.thehopnerd.com

Ian Allison is a husband, father, and a friend. He is a Certified Safety Professional who has worked in the safety field for a decade. He has a Bachelor of Arts in Environmental Studies and Native American Studies from Dartmouth College and a Master of Business Administration from Arizona State University. He has worked primarily in power generation, supply chain, and public universities. He is from Tuba City, Arizona on the Navajo Nation. He currently resides in Phoenix, Arizona with his wife, Beth, and son, Cornell. He loves spending time with his family, fishing, running, and consuming podcasts. He is also the host of his own podcast, Native Film Talk. A podcast discussing Native American representation in film and television. His mantras in life are "Never stop learning," and "Never be afraid to say, 'I don't know.'"

For more about Ian, visit www.nativefilmtalk.com

For More...

Tune into *The HOP Nerd Podcast*, available everywhere you listen to podcasts, for weekly in-depth conversations about doing *safety better*.

For training, consulting, help, and support in all thing's *safety better*, visit www.safetybetterment.com or www.thehopnerd.com.

This book has been brought to you by

Visit www.palehorsemedia.co for books, podcasts, and more!

www.ingramcontent.com/pod-product-compliance
Lightning Source LLC
Chambersburg PA
CBHW060828220526
45466CB00003B/1023